T0269277

Cancer Treatment and the Ovary

Cancer Treatment and the Ovary

Clinical and Laboratory Analysis of Ovarian Toxicity

Edited by

Richard A. Anderson

Norah Spears

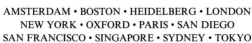
AMSTERDAM • BOSTON • HEIDELBERG • LONDON
NEW YORK • OXFORD • PARIS • SAN DIEGO
SAN FRANCISCO • SINGAPORE • SYDNEY • TOKYO

Academic Press is an imprint of Elsevier

Academic Press is an imprint of Elsevier
125, London Wall, EC2Y 5AS.
525 B Street, Suite 1800, San Diego, CA 92101-4495, USA
225 Wyman Street, Waltham, MA 02451, USA
The Boulevard, Langford Lane, Kidlington, Oxford OX5 1GB, UK

ISBN: 978-0-12-801591-9

Library of Congress Cataloging-in-Publication Data
A catalog record for this book is available from the Library of Congress.

British Library Cataloguing-in-Publication Data
A catalogue record for this book is available from the British Library.

For Information on all Academic Press publications
visit our website at http://store.elsevier.com/

ELSEVIER   Book Aid International   Working together to grow libraries in developing countries

www.elsevier.com • www.bookaid.org

CONTENTS

LIST OF CONTRIBUTORS

Richard A. Anderson
MRC Centre for Reproductive Health, University of Edinburgh, Edinburgh, UK

Claus Yding Andersen
Laboratory of Reproductive Biology, Copenhagen University Hospital –
Rigshospitalet, University of Copenhagen, Copenhagen, Denmark

Marilia Henriques Cordeiro
Department of Obstetrics and Gynecology, Feinberg School of Medicine,
Northwestern University, Chicago, IL, USA

Hakan Cakmak
Division of Reproductive Endocrinology and Infertility, Department of Obstetrics,
Gynecology, and Reproductive Sciences, University of California, San Francisco,
CA, USA

Christopher H. Grant
Institute of Genetics and Molecular Medicine, University of Edinburgh Cancer
Research UK Centre, Western General Hospital, Edinburgh, UK

Charlie Gourley
Institute of Genetics and Molecular Medicine, University of Edinburgh Cancer
Research UK Centre, Western General Hospital, Edinburgh, UK

Annette Klüver Jensen
Laboratory of Reproductive Biology, Copenhagen University Hospital –
Rigshospitalet, University of Copenhagen, Copenhagen, Denmark

So-Youn Kim
Department of Obstetrics and Gynecology, Feinberg School of Medicine,
Northwestern University, Chicago, IL, USA

Stine Gry Kristensen
Laboratory of Reproductive Biology, Copenhagen University Hospital –
Rigshospitalet, University of Copenhagen, Copenhagen, Denmark

Stephanie Morgan
Centre for Integrative Physiology, University of Edinburgh, Edinburgh, UK

Dror Meirow
Center for Fertility Preservation, Sheba Medical Center, Tel Hashomer, Israel

Kutluk Oktay
Division of Reproductive Medicine and Laboratory of Molecular Reproduction &
Fertility Preservation, Obstetrics and Gynecology, New York Medical College,
Valhalla, NY, USA; Innovation Institute for Fertility Preservation and IVF,
New York, NY, USA

Elon C. Roti Roti
Department of Medicine, University of Wisconsin, Madison, WI, USA

Mitchell P. Rosen
Division of Reproductive Endocrinology and Infertility, Department of Obstetrics, Gynecology, and Reproductive Sciences, University of California, San Francisco, CA, USA

Hadassa Roness
Center for Fertility Preservation, Sheba Medical Center, Tel Hashomer, Israel

Sana M. Salih
Department of Obstetrics and Gynecology, University of Wisconsin, Madison, WI, USA

Norah Spears
Centre for Integrative Physiology, University of Edinburgh, Edinburgh, UK

Volkan Turan
Division of Reproductive Medicine and Laboratory of Molecular Reproduction & Fertility Preservation, Obstetrics and Gynecology, New York Medical College, Valhalla, NY, USA; Innovation Institute for Fertility Preservation and IVF, New York, NY, USA

Teresa K. Woodruff
Department of Obstetrics and Gynecology, Feinberg School of Medicine, Northwestern University, Chicago, IL, USA

Mary B. Zelinski
Division of Reproductive & Developmental Sciences, Oregon National Primate Research Center, Oregon Health & Science University, Beaverton, OR, USA; Department of Obstetrics & Gynecology, Oregon Health & Science University, Portland, OR, USA

FOREWORD

Look deep into nature, and then you will understand everything better.

Albert Einstein

The last five decades have seen a significant improvement in the survival rate following the treatment of children and young people with cancer. In the less enlightened days of the 1950's and early 1960's leukaemia in children was considered to be incurable. As a direct consequence the pioneering physicians who first treated children with leukaemia with chemotheraputic agents were initially ignored and ridiculed. Currently around 80% of children and young people with cancer will be alive five years after diagnosis, and for acute lymphoblastic leukaemia the cure rate is currently in excess of 90%.

With increasing survival and cure rates our attention has turned more towards the quality of the cure for these young people. Perhaps the most common questions our patients ask once the fear of relapse has diminished are am I fertile and, if I am able to have children will my children be normal? To help answer these questions Richard Anderson and Norah Spears have brought together a multidisciplinary group of experts on the effect of cancer treatment on the ovary. The focus is both clinical and lab-based and takes us on a journey from the laboratory to the bedside and back again.

One of the greatest challenges when faced with a new young patient with cancer is to explain to them not only their diagnosis and treatment plan but also what are the short-, medium- and long-term side effects of the treatment. The importance of addressing fertility and ovarian function in young woman with cancer is twofold. Firstly, bringing up the subject and talking about future fertility gives hope to the young patient and their family that the team looking after them believe that they will survive, and that issues of ovarian function and fertility are going to be of importance to them in the future. Secondly, if ovarian function and fertility are likely to be compromised by the patient's treatment, consideration should be given to fertility preservation strategies before treatment commences. This presents a huge challenge for the multidisciplinary team at a time of immense stress and upheaval for the young patient and their family.

As our knowledge of the effects of chemotherapy and radiotherapy on ovarian function has expanded, our understanding of normal ovarian function has blossomed. In this book the editors have brought together experts from diverse scientific and clinical backgrounds to explain normal ovarian function, the effect of chemotherapy on ovarian function, and techniques both established and experimental that may preserve ovarian function in a young patient with cancer.

This book is timely, and the issues are discussed in a lucid, balanced and provocative manner. From the debate about the presence and functionality of stem cells within the ovary, to the potential for ovarian protection from GnRH analogues, this book provides a rich recourse not only for those interested in understanding more about normal ovarian function but also for those responsible for looking after young female patients with cancer. This book will be of interest not only to reproductive endocrinologists and scientists interested in the normal physiology of ovarian function but also to oncologists committed to improving the quality of the cure for their young patients.

I enjoyed this book and I would recommend it to you as an ever present companion as it seeks to explain in a clear and comprehensive manner the effect of cancer treatment on ovarian function and what, if anything, can be done to potentially restore ovarian function and fertility in the future.

Professor W. Hamish B Wallace
Professor of Paediatric Oncology, Royal Hospital for Sick Children, Edinburgh & The University of Edinburgh.

INTRODUCTION

This book aims to bring together a comprehensive review and analysis of our understanding of the effects of cancer treatment on the ovary. The idea behind the book stems from the growing importance of retaining reproductive function in girls and young women after cancer treatment, reflecting increasing survival: the emphasis is now to optimise post cancer life, and minimise the adverse effects of treatment on health. The last decade has seen an explosion in clinical interest in this field, particularly with the development of oocyte vitrification and ovarian tissue cryopreservation as strategies to preserve fertility in girls and young women facing a cancer diagnosis. Here, we bring together a range of clinical and basic science chapters to provide a comprehensive review of our current understanding of the effects of cancer therapies on the ovary, and of potential strategies to protect against gonadotoxicity.

The first Section of the book focuses on the current clinical situation. In Chapter 1, Teresa Woodruff and colleagues provide a comprehensive review of the relevant female physiology, from follicle formation during fetal life through follicle development, to ovulation and fertilisation. This provides a foundation for all subsequent chapters, highlighting key vulnerabilities of this system at various stages. Chapter 2 brings a more focused oncological perspective, with Christopher Grant and Charlie Gourley describing the most common chemotherapeutic agents administered to girls and young women, along with their main actions on the ovary, where known: this chapter also serves to highlight the many gaps in our understanding of the effects of these agents on the ovary, particularly when given in combination as is normally the case. In Chapter 3, the clinical perspective is continued by Richard Anderson with a discussion of how gonadal damage can be assessed clinically using current biomarkers of ovarian function. This is further detailed in Chapter 4, where Volkan Turan and Kutluk Oktay provide a more detailed description of some of the key chemotherapeutic regimens used, and their effects on fertility in the relevant common diseases in reproductively aged women, highlighting new understanding of the role of DNA damage repair mechanisms in this process.

Greater understanding of clinical effects of cancer treatment require the use of experimental models and Chapters 5 and 6 address this area, with this second section examining the available laboratory models. In Chapter 5, Mary Zelinski and colleagues outline current data from *in vivo* models, both rodents and non-human primates, also introducing the recent focus on trying to protect the ovary against the adverse effects of chemotherapy. *In vitro* approaches are then described in Chapter 6, where Stephanie Morgan and Norah Spears highlight the range of culture methods that have been used, culturing cells, follicles and ovary from humans as well as rodents and non-human primates. Perhaps the overriding message from both these chapters, though, is that there has been surprisingly little use of such experimental models, with much more such work badly needed.

The final section returns to a clinical perspective, focussing on strategies aiming to protect the ovaries of girls and young women with cancer during treatment, covering both those in use and those currently under development. Claus Yding Andersen and colleagues review the current status of ovarian tissue cryopreservation in Chapter 7, while in Chapter 8, Hakan Cakmak and Mitch Rosen provide a comprehensive review of the use of hormonal approaches, particularly that of GnRH analogues, examining their ability to protect the ovary during chemotherapy treatment. The final chapter of the book, Chapter 9, turns to a preclinical perspective, with Hadassa Roness and Dror Meirow reviewing the various approaches currently under investigation for their ability to reduce the adverse effects of chemotherapy on the ovary.

With an ever-growing population of cancer survivors, the effect of cancer treatment on subsequent fertility is an important area, and will become only more of a focus in the decades to come. There is a growing literature in this area, but we believe that by bringing the various clinical and non-clinical aspects together into one publication, this book can provide a much needed reference point for the field and highlight the major gaps in our understanding.

Our thanks to Molly McLaughlin and others at Elsevier for support throughout this venture, and to the Medical Research Council for funding our own research in this area. We would also like to thank Ronnie Grant for help with illustrations, including for the book cover. Finally, though, we are grateful to all the authors who have contributed to the book, which we believe brings together the various aspects of this field in a timely and comprehensive way.

Richard A. Anderson and Norah Spears

SECTION I

Clinical

CHAPTER 1

Ovarian Follicle Biology and the Basis for Gonadotoxicity

Marilia Henriques Cordeiro, So-Youn Kim, and Teresa K. Woodruff

Department of Obstetrics and Gynecology, Feinberg School of Medicine, Northwestern University, Chicago, IL, USA

1.1 OVERVIEW OF OVARIAN FUNCTION

The ovary serves two roles — the production of hormones necessary to support the endocrine health of the individual and the generation of mature oocytes that are able to be fertilized and contribute half of the genetic makeup of a new organism. The ovarian follicle is the functional unit of the ovary that carries out both of these goals.[1,2] The ovarian follicles within an adult ovary produce steroid and protein hormones in cyclical patterns in response to tropic factors from the hypothalamus and pituitary (Figure 1.1). These hormones then act locally to support follicle and oocyte growth and development, as well as systemically to support reproductive health, bone health and cardiovascular function.[3] Interruption of the endocrine hormones at the level of the brain or gonad due to cancer treatment can result in loss of cyclicity and diminished endocrine health for short periods of time or can render an individual sterile due to loss of ovarian follicles.

Each ovarian follicle consists of one oocyte surrounded by supporting somatic cells, the granulosa and theca cells,[4] and the follicle structure is embedded in an ovarian stroma that provides additional critical signals that determine follicle fate.[5] Follicle development starts with selection of follicles from a pool of dormant primordial follicles — the ovarian reserve. The recruited follicles are activated and undergo a series of changes in morphology and size, developing through primary and secondary stages before acquiring a fluid-filled antral cavity.[4] Although multiple follicles are selected to begin growing and eventually reach the antral stage, many are lost to atresia and only one (or a few depending on the species) reaches the Graffian follicle stage.[1,6]

Cancer Treatment and the Ovary. DOI: http://dx.doi.org/10.1016/B978-0-12-801591-9.00001-1

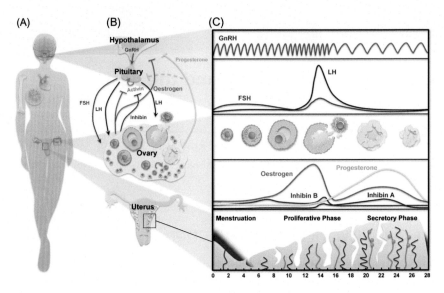

Figure 1.1 **Ovarian endocrine control of reproductive cyclicity and female systemic health.** *(A) Ovarian-derived hormones are important to maintain normal homeostasis of peripheral organs, including brain, heart, mammary glands, bones, reproductive tract, etc. Therefore, systemic health complications are associated with interruption of ovarian function at menopause or due to anticancer treatment-induced premature ovarian failure (POF). The hypothalamus–pituitary–ovary axis is represented in blue. (B) Normal ovarian function and cyclicity are regulated by a complex hormonal feedback loop between the organs of the hypothalamus–pituitary–ovary axis. Hypothalamic gonadotropin-releasing hormone (GnRH) pulses control pituitary gonadotropin (luteinizing hormone [LH] and follicle-stimulating hormone [FSH]) release, which regulates follicle growth and production of steroid and peptide hormones (oestrogen, inhibins and progesterone). These gonadal hormones then exert a negative feedback on the pituitary and hypothalamus. The frequency of GnRH pulses determines FSH and LH release, stimulating follicle development. During follicular growth oestrogen and inhibin are produced, causing negative feedback in the hypothalamus and pituitary. Increased frequency of GnRH pulses induces LH release, triggering ovulation. The residual follicle structure forms the corpus luteum and produces progesterone, which is responsible for preparing the endometrium (uterine wall) for implantation and pregnancy. If pregnancy is not achieved, the uterine wall disintegrates causing the menses, at a time when a new group of gonadotropin-responsive follicles will start to grow, initiating a new cycle. The hormonal levels and fluctuations have been simplified to better represent the overall secretion pattern throughout the menstrual cycle.*

Ovulation occurs in response to a timed surge of pituitary gonadotropins (see Figure 1.1) that triggers the release of a mature oocyte from the surface of the ovary into the fallopian tube, where it can be fertilized. The residual follicle unit (the remaining theca and mural granulosa cells) undergoes transformation to become the corpus luteum.

Ovarian follicle development is a carefully orchestrated event that is directed by both endocrine signalling within the hypothalamic–pituitary–ovarian axis as well as paracrine signalling occurring within and between follicles in the ovary.[7] Hormone-regulated follicular development involves the integration of feedback loops between oestrogen, progesterone, inhibin A and inhibin B from the ovary;

follicle-stimulating hormone (FSH) and luteinizing hormone (LH) from the pituitary; and hypothalamic gonadotropin-releasing hormone (GnRH) to direct follicle and oocyte development (see Figure 1.1).[8] On the other hand, early follicle development is driven by intrinsic ovarian factors[9] that regulate oocytic and follicular development before the follicles acquire FSH responsiveness.[10-13] These locally acting factors and their roles in maintaining the dormant follicle pool, activating selected primordial follicles and supporting the earliest stages of follicle development are less well understood and of intense interest in the reproductive field. Understanding the early signalling events associated with maintaining a health-quiescent follicle pool may provide insights into gonadotoxicity and how to mitigate this effect.

1.2 OVARIAN DEVELOPMENT

The mammalian gonad is formed during embryonic development (Figure 1.2) after migration of primordial germ cells (PGCs) from the hindgut into the genital ridge between 9 to 10.5 days post coitum (dpc) in the mouse and at 6 weeks of gestation in humans.[14-16] During this process, PGCs undergo mitotic divisions[17] with incomplete cytokinesis resulting in cysts where germ cells remain connected by cytoplasmic bridges.[18] At this stage, the gonad is bipotent. In the presence of a Y chromosome, the expression of the *Sry* gene (sex-determining region on the Y chromosome) directs male development through a process called sex determination.[19] In contrast, sex determination in females requires the activity of several genes, such as *Wnt4*, *Rspo1*, *β-Catenin*, *Foxl2* and *Fst*. Loss of these genes during development causes various degrees of female-to-male sex reversal.[20,21]

Following female-specific gene activation, the early ovary develops distinct structural features, including the formation of ovigerous cords by interaction of the germ cell clusters with surrounding pregranulosa cells.[22,23] Around 13.5 dpc, mouse germ cells enter meiosis in an anterior-to-posterior gradient believed to be driven by retinoic acid (RA)[24-27]; in the human ovary, this process begins in the medulla and spreads radially into the cortex from 10 weeks gestation.[16,28,29] Reductive division (meiosis) begins during embryonic development in the female gonad, and postnatally in the male germ cell, respectively. Although the developing testis is also exposed to RA, the production of CYP26B1 allows degradation of RA and the male germ cells remain

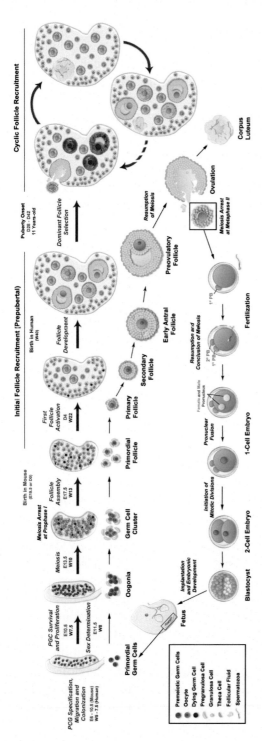

Figure 1.2 **Life cycle of the female germ cell.** The primordial follicles originate from primordial germ cells (PGCs) after their migration into the genital ridge, proliferation, sex determination, meiosis and individual follicle assembly. Individual primordial oocytes are recruited to grow and produce mature oocytes through the process of folliculogenesis. The oocyte increases in size as the number and complexity of the granulosa and theca cell layers expand. When the follicle reaches the antral stage, it forms a fluid-filled cavity, the antrum, causing differentiation of the granulosa cell layers into mural and cumulus cells. Following the luteinizing hormone (LH) surge, the oocyte resumes meiosis becoming arrested at metaphase II (MII), while cumulus cells expand and the follicle wall is prepared for ovulation. The MII oocyte is then released from the follicle cavity, ovulated, and if fertilized by available sperm, becomes a zygote and initiates embryogenesis. The embryo then travels through the fallopian tube and arrives at the uterus where implantation takes place. During embryogenesis, the presumptive germ cells are established and ultimately move to the developing bipotent gonad where the oogonia life cycle begins again. The initial and cyclic waves of follicular growth are also represented. Embryonic germ cell loss by attrition and follicular degeneration by atresia are represented by dark purple and dark green, respectively. The timing of events is approximated and represented in embryonic days (E) for mouse (green) and weeks of gestation (W) for human (blue), with E19.5 and W40 being the day of birth, respectively.

arrested in G1/G0.[26,30] Oogonia enter the first stages of meiosis and arrest in diplotene of prophase I; at this time, they are referred to as oocytes and remain arrested in this stage until the LH surge that triggers ovulation, with the final stages of meiotic division completed upon fertilization.

The newly formed oocytes become individually encapsulated in a single layer of somatic cells to form primordial follicles (see Figure 1.2). In contrast to humans and other large mammals, most primordial follicles in rodents are formed a few days after birth, with some being activated soon after formation.[1] This process of primordial follicle formation is also called nest breakdown because the pregranulosa cells invade and degrade the cytoplasmic bridges within the cysts while some germ cells are systematically eliminated.[18] Several factors have been identified that control the timing and extent of nest breakdown, including *Figla*, *Foxl2*, *Nobox*, *c-Kit* (oocyte)/*Kit* ligand (pregranulosa and granulosa cells), members of the transforming growth factor beta (TGFβ) superfamily, oestrogens and phytoestrogens, progesterone, extracellular matrix factors, neurotrophins, and members of the *Notch* signalling pathway.[31,32]

Any insults that occur during ovarian embryonic development may negatively impact adult ovarian function; these include poor nutrition, exposure to environmental pollutants, and exposure to certain drugs.[33] Thus, ovarian development represents a critical window of vulnerability that can influence normal gonadal function in the adult animal.

1.3 MOLECULAR MECHANISMS CONTROLLING PRIMORDIAL FOLLICLE ACTIVATION

The primordial follicles are the first class of follicles formed in mammalian ovaries and consist of an oocyte surrounded by a single layer of flattened granulosa cells. These follicles remain in a dormant state until they receive signals for activation or are relieved of a negative regulatory factor. Dormancy is characterized by oocytic meiosis arrest at prophase I, pregranulosa G1/G0 mitotic arrest,[34,35] and a relatively cortical location within the ovary with little or no vascular access.[36] Each primordial follicle has four possible fates: remain quiescent, die by attrition, begin development but later be lost by atresia, or begin development and ultimately release an oocyte followed by formation of a short-lived corpus luteum (see Figure 1.2).[37–39] A small number of primordial follicles are continuously activated to exit the ovarian reserve and begin growing. When development begins, the oocyte expands in size as metabolic activity increases and the surrounding pregranulosa cells start to proliferate and differentiate.[1] Since follicle activation is irreversible, and the starting pool of follicles resulting from nest breakdown is finite, follicle activation is tightly controlled to permit continuous waves of follicle growth during the ~40 years of the female reproductive lifespan.[38,40] It is believed that the trigger(s) for primordial follicle activation are intraovarian signals,[9] but it is still unknown if signals originate in the oocyte or the granulosa cells. Many studies have identified important molecular players controlling global primordial follicle activation; however, the mechanisms underlying selective follicular activation over decades of life – namely, how certain primordial follicles enter the growth phase while others remain quiescent – are still largely unknown. One popular explanation for the order of follicle activation is the "production line" hypothesis. This hypothesis suggests that the first oogonia to undergo meiotic arrest are the first to become activated, implying that the order of follicle activation is established during embryonic development.[41] However, this hypothesis remains controversial, with evidence both in favour[34,42,43] and against.[44,45]

Early studies showed that primordial follicles spontaneously activate when cultured *in vitro*, suggesting that quiescence is based on inhibitory factors present *in vivo*, although a stimulatory effect of factors in the culture medium could not be excluded.[46] Indeed, a number of growth factors have been shown to be important for primordial follicle recruitment and maintenance, including vascular endothelial growth factor (VEGF), leukaemia inhibitor factor (LIF), basic fibroblast growth factor (bFGF), keratinocyte growth factor (KGF), platelet-derived growth factor (PDGF), neurotrophins, bone morphogenetic protein-4 (BMP-4), and BMP-7 and c-Kit/Kit ligand.[47]

In vitro studies have shown that c-Kit/Kit ligand signalling activates the PI3K signalling network in oocytes, promoting their survival and growth.[48] Multiple studies have demonstrated the importance of the PI3K–PTEN–AKT–mTOR signalling pathway in controlling primordial follicle dormancy.[2,49] Transgenic animals in which *Foxo3*, *Tsc1* and *Tsc2*, *PTEN*, and *p27kip1* have been genetically ablated undergo premature ovarian failure (POF) due to global follicle activation.[49] Genetic depletion of *PDK1* (phosphoinositide-dependent kinase 1) and

RpS6 (ribosomal protein 6) also causes POF, although it occurs through accelerated follicle clearance and atresia,[50] suggesting that a certain level of PDK1 expression is important to control the balance between primordial follicle activation and dormancy. Manipulation of the PI3K pathway has been studied as an approach to induce activation of mouse and human primordial follicles *in vitro*[51−53]; however, a recent study noted detrimental effects of this strategy on follicle development[54] and further safety studies will be required before application to humans. Taken together, these studies indicate that the PI3K signalling pathway is a critical determinant of primordial follicle activation and survival.

Other transcriptional factors also contribute to follicle activation. Forkhead Box L2 (*FoxL2*) expression in granulosa cells has been implicated in primordial follicle activation by inducing the transition from squamous to cuboidal granulosa cells. Despite delayed nest breakdown, *FoxL2* knockout mice form individual primordial follicles, which then arrest and undergo atresia at early stages of development, resulting in premature ovarian failure by 2 months of age.[55,56] Blockage at the primordial follicle stage and accelerated germ cell loss have also been observed after genetic deletion of *Nobox*,[57] *Lhx8*, *Sohlh-1* (spermatogenesis and oogenesis helix-loop-helix 1) and *Sohlh-2*.[58,59]

Others have proposed that substances produced by growing follicles, including anti-Müllerian hormone (AMH; also called Müllerian inhibitory substance [MIS]), exert an inhibitory influence on primordial oocytes, keeping them in a dormant state.[60] AMH is also a useful biomarker of growing follicles[61,62] and can be used for assessment of the ovarian reserve in patients following treatment for cancer[63]; this is reviewed in Chapter 3. AMH produced by the granulosa cells of growing follicles[64] appears to inhibit primordial follicle recruitment, as mice lacking AMH have an elevated number of growing follicles and show depletion of the primordial follicle pool earlier than their wild-type littermates.[60] Other factors, such as the neuropeptide somatostatin (SST) and the chemoattractive cytokine CXCL12, are also known to inhibit the recruitment of primordial follicles.[65,66]

In addition to the lack of understanding of the mechanisms controlling selective follicle recruitment, the extent and timing of the first wave of follicle activation (first growing follicles in the prepubertal ovary) are still largely unknown. Interestingly, recent work suggests that the timing of granulosa cell specification may contribute to fundamental differences between the initial and subsequent cyclical waves of primordial follicle recruitment and the maintenance of primordial oocyte quiescence.[35] The first wave of growing follicles reaches the antral stage before puberty; in the absence of FSH, these follicles are believed to be unable to develop further and therefore lost to atresia.[67] However, a recent study in mice challenged this concept by showing that these first-wave follicles not only induce the onset of puberty but also grow faster than follicles in the adult ovary and are ovulated.[68] Further studies focusing on the similarities and differences between first and consecutive waves of primordial follicle activation are important to understand the molecular mechanisms controlling selective follicle recruitment.

1.4 LOCAL CONTROL OF EARLY FOLLICLE DEVELOPMENT

After a primordial follicle has been activated and growth is initiated, the fate of the follicle is controlled by various interacting endocrine, paracrine, and autocrine factors.[7,69,70] Preantral follicle growth is gonadotropin independent, as revealed in mutant mice with compromised FSH signalling that showed follicle development arrested only prior to antrum formation.[10–13] The rate of early follicular development is dependent on the oocyte,[7,70] which produces several growth factors, such as GDF9 and BMP15, that act on the granulosa cells. Follicles from mice null for these growth factors fail to progress to later stages of development.[71,72] Conversely, overexpression of BMP15 causes an early decline in the ovarian reserve by accelerating follicle growth, suggesting that BMP15 is important for controlling follicle growth and preventing maturation.[73] Bidirectional communication between the oocyte and its surrounding somatic cells plays a critical role in follicular development.[7,69] Gap junctions between the oocyte and the supporting granulosa cells facilitate an exchange of signalling molecules during folliculogenesis. Deletion of connexin 37 (oocyte–granulosa cell gap junctions) and connexin 43 (granulosa–granulosa cell gap junctions) causes defects in oocyte meiotic competence and follicle growth arrest at the primary stage, respectively.[74–76] Other factors have been shown to be important for early follicle growth, such as cyclin D2,[77] IGF-1,[78] and TAF4B.[79–81] Additionally, formation of the theca cell layer is a crucial feature of early follicle growth.[82] This cell type is believed to be differentiated from mesenchymal precursors present in the ovarian stroma[83] and is the primary source for androgen production.[84]

1.5 ENDOCRINE CONTROL OF LATER FOLLICLE DEVELOPMENT

During the preantral to antral transition, scattered fluid-filled spaces between granulosa cells converge to form a single antral cavity (see Figures 1.1 and 1.2). Accumulation of follicular fluid causes a rapid expansion in follicle size, while the theca layer differentiates into theca interna (steroidogenic activity) and theca externa (highly vascularized containing macrophages, smooth muscle-like cells and fibroblasts) compartments.[8,84,85] The formation of the antrum also marks the differentiation of the granulosa cell layer into mural granulosa cells and cumulus cells. Antral follicles become receptive to extraovarian regulation and endocrine signalling within the hypothalamic–pituitary–gonadal (HPG) axis. The hypothalamus produces and releases pulses of GnRH into the pituitary blood supply, inducing pituitary release of FSH and LH in response to low- or high-frequency pulses, respectively (see Figure 1.1). FSH is essential for granulosa cell proliferation, oestradiol production, LH receptor expression and the prevention of granulosa cell apoptosis (atresia).[13,86–88] Follicular oestrogens and inhibins exert negative feedback on the HPG axis that ultimately triggers an LH surge.[8,89,90] During each wave of folliculogenesis, most of the recruited follicles are lost to atresia at the antral stage, while only a few (or one in humans) grow further to reach the preovulatory stage.[1,6] When the follicle reaches the final stages of maturation, the LH surge induces meiotic resumption in the oocyte, breakdown of the follicle wall and extrusion of the cumulus–oocyte complex by ovulation.[85] The remaining granulosa cells in the postovulatory follicle undergo luteinization and form a corpus luteum. This transient endocrine organ produces progesterone, which is essential for maintenance of pregnancy.[91,92] Any insult affecting the activity of the HPG axis can compromise the oestrus/menstrual cycle and cause severe disruption of folliculogenesis and infertility.

1.6 REPRODUCTIVE LIFESPAN AND THE OVARIAN PRIMORDIAL FOLLICLE RESERVE

Fertility is determined by the number of primordial follicles available in the ovary, the ovarian reserve, formed upon nest breakdown during human ovarian embryonic development. There is contradictory evidence suggesting that neo-oogenesis occurs in the postnatal mammalian ovary,[93,94] but it is generally accepted that the mammalian ovary contains a finite number of primordial oocytes, and that this source declines with age.[1,40,95,96] When the available primordial follicle pool is depleted, the reproductive capacity of a woman ceases; thus, the ovarian reserve is a measure of reproductive lifespan. In humans, the total number of primordial follicles falls drastically from a peak of 6−7 million at 20 weeks of gestation, to approximately 1−2 million viable oocytes in early neonatal life.[40,97] At the time of menopause, the number of follicles remaining in the ovarian reserve drops below 1000. Due to an exhaustion of the ovarian follicle reserve, menopause occurs at about 51 years of age, a time point that has been constant for centuries.[95] Premature ovarian failure (POF), also known as primary ovarian insufficiency (POI) or premature menopause, refers to cessation of ovarian function before the age of 40. This condition affects 1% of the female population and is idiopathic in the majority of cases; the remaining cases of POI have been linked to chromosomal, genetic, metabolic, autoimmune and infectious causes.[98] POI can also occur as a side effect from anticancer treatment, which is discussed in detail in Chapters 3 and 4. A delicate balance between primordial follicle activation and loss is required to prevent premature depletion of the ovarian reserve as well as to protect against health complications caused by a loss of ovarian hormones, which have systemic effects on various tissues, including bone and the cardiovascular system. Understanding the molecular mechanisms involved in primordial follicle activation and loss is crucial if we are to intervene to maintain reproductive lifespan and endocrine function.

1.7 GERM CELL SENSITIVITY TO CELL DEATH

As previously mentioned, primordial oocytes arrested at prophase I can remain quiescent for many years before being recruited. Prolonged arrest combined with the unique status of the oocyte nucleus makes primordial oocytes particularly vulnerable to DNA damage caused by environmental stress or anticancer therapy.[99] In fact, small preantral follicles are easily lost by chemotherapy and radiation therapy, while larger preantral follicles seem more resistant to the same dosage of the treatment.[100,101] It has been suggested that the sensitivity of small follicles to anticancer treatment is associated with their high endogenous nuclear expression of TAp63.[101] To avoid carrying mutations to the next generation, the oocyte detects and repairs DNA damage or undergoes programmed cell death when genomic integrity becomes overly compromised.[102,103] As a consequence, oocyte loss by apoptosis can quickly lead to ovarian reserve depletion and infertility.[104,105] Understanding the molecular mechanisms controlling oocyte death after DNA damage is imperative to prevent treatment-induced POI. Oocyte manipulation to increase the resistance to DNA damage-induced apoptosis carries the risk of promoting the survival of oocytes carrying potential mutations; however, it is also possible that given sufficient time, a damaged oocyte that is prevented from undergoing apoptosis may be able to repair the DNA damage.[106] Even if the DNA damage cannot be reversed, stopping the uncontrolled elimination of the ovarian reserve would be beneficial to maintain ovarian endocrine activity and avoid health complications associated with early menopause.

Interestingly, cell degeneration occurs extensively in a normal ovary: attrition/degeneration of germ cells occurs during embryonic development, which accounts for the largest loss of oocytes, and atresia/follicle degeneration occurs postnatally with each wave of follicle development during reproductive life.[6,107] Apoptosis,[18,108–111] autophagy,[37,39,111] and direct exclusion from the ovaries[39] are the death mechanisms that contribute to germ cell attrition. It has been hypothesized that germ cell loss may be a mechanism in place to ensure oocyte quality, allowing elimination of germ cells with damaged nuclei or organelles and ensuring genomic integrity.[18,38] Another common explanation for germ cell loss is to ensure an appropriate ratio between germ cells and supporting cells upon nest breakdown.[112] Molecular players important to the regulation of germ cell loss and survival include the anti-apoptotic gene *Bcl-2*, deletion of which reduces oocytes and primordial follicles.[113] Deletion of the pro-apoptotic factor *Bax* results in an increased number of primordial oocytes and prolonged reproductive lifespan.[114–116] Moreover, deletion of *Caspase-2*,[117] *TAp63*,[118,119] *Puma* and *Puma/Noxa*[106] results in an increase in the number of primordial follicles, as well as increased resistance to radiation or chemotherapeutic agents. Deletion of *Zfx* results in accelerated germ cell loss at birth as well as postnatal germ cell loss and POI; however, the mechanism of action of this factor remains unknown.[120] Future studies are required to further illuminate the molecular mechanisms specific to germ cell death and how these can be manipulated to develop safe approaches to prevent ovarian reserve depletion during gonadotoxic treatment; this approach is discussed in detail in Chapter 9. In parallel, the development of new chemotherapeutic drugs and delivery systems will be vital to minimize toxicity and off-target detrimental effects on fertility and endocrine health.[121]

1.8 CONCLUSION

In summary, the primordial follicle pool present at birth represents a woman's entire reproductive potential throughout her lifespan. The intricate interplay of growth factors and hormones that govern whether a particular follicle remains dormant, undergoes atresia or is activated to grow and ovulate is a remarkable process and the biology is slowly being revealed. Understanding the molecular mechanisms that allow communication between the germ cells and somatic support cells that dictate ovarian development and follicle loss and that either maintain oocyte quiescence or drive activation will be critical to finding ways to understand and protect the ovarian reserve, and thereby the reproductive health of women.

REFERENCES

1. McGee EA, Hsueh AJ. Initial and cyclic recruitment of ovarian follicles. *Endocr Rev.* 2000;21(2):200–214.

2. Adhikari D, Liu K. Molecular mechanisms underlying the activation of mammalian primordial follicles. *Endocr Rev.* 2009;30(5):438–464.

3. Nejat EJ, Chervenak JL. The continuum of ovarian aging and clinicopathologies associated with the menopausal transition. *Maturitas.* 2010;66(2):187–190.

4. Peters H, Byskov AG, Himelstein-Braw R, Faber M. Follicular growth: the basic event in the mouse and human ovary. *J Reprod Fertil.* 1975;45(3):559–566.

5. Berkholtz CB, Shea LD, Woodruff TK. Extracellular matrix functions in follicle maturation. *Semin Reprod Med.* 2006;24(4):262–269.

6. Hsueh AJ, Billig H, Tsafriri A. Ovarian follicle atresia: a hormonally controlled apoptotic process. *Endocr Rev.* 1994;15(6):707–724.

7. Eppig JJ. Oocyte control of ovarian follicular development and function in mammals. *Reproduction.* 2001;122(6):829–838.

8. Edson MA, Nagaraja AK, Matzuk MM. The mammalian ovary from genesis to revelation. *Endocr Rev.* 2009;30(6):624–712.

9. Edwards RG, Fowler RE, Gore-Langton RE, et al. Normal and abnormal follicular growth in mouse, rat and human ovaries. *J Reprod Fertil.* 1977;51(1):237–263.

10. Cattanach BM, Iddon CA, Charlton HM, Chiappa SA, Fink G. Gonadotrophin-releasing hormone deficiency in a mutant mouse with hypogonadism. *Nature.* 1977;269 (5626):338–340.

11. Dierich A, Sairam MR, Monaco L, et al. Impairing follicle-stimulating hormone (FSH) signalling in *vivo*: targeted disruption of the FSH receptor leads to aberrant gametogenesis and hormonal imbalance. *Proc Natl Acad Sci USA*. 1998;95(23):13612–13617.

12. Abel MH, Wootton AN, Wilkins V, Huhtaniemi I, Knight PG, Charlton HM. The effect of a null mutation in the follicle-stimulating hormone receptor gene on mouse reproduction. *Endocrinology*. 2000;141(5):1795–1803.

13. Kumar TR, Wang Y, Lu N, Matzuk MM. Follicle stimulating hormone is required for ovarian follicle maturation but not male fertility. *Nat Genet*. 1997;15(2):201–204.

14. Anderson R, Copeland TK, Scholer H, Heasman J, Wylie C. The onset of germ cell migration in the mouse embryo. *Mech Dev*. 2000;91(1–2):61–68.

15. Molyneaux KA, Stallock J, Schaible K, Wylie C. Time-lapse analysis of living mouse germ cell migration. *Dev Biol*. 2001;240(2):488–498.

16. Hartshorne GM, Lyrakou S, Hamoda H, Oloto E, Ghafari F. Oogenesis and cell death in human prenatal ovaries: what are the criteria for oocyte selection? *Mol Hum Reprod*. 2009;15(12):805–819.

17. Tam PP, Snow MH. Proliferation and migration of primordial germ cells during compensatory growth in mouse embryos. *J Embryol Exp Morphol*. 1981;64:133–147.

18. Pepling ME, Spradling AC. Mouse ovarian germ cell cysts undergo programmed breakdown to form primordial follicles. *Dev Biol*. 2001;234(2):339–351.

19. Koopman P, Gubbay J, Vivian N, Goodfellow P, Lovell-Badge R. Male development of chromosomally female mice transgenic for Sry. *Nature*. 1991;351(6322):117–121.

20. Maatouk DM, Capel B. Sexual development of the soma in the mouse. *Curr Top Dev Biol*. 2008;83:151–183.

21. Ottolenghi C, Pelosi E, Tran J, et al. Loss of Wnt4 and Foxl2 leads to female-to-male sex reversal extending to germ cells. *Hum Mol Genet*. 2007;16(23):2795–2804.

22. Konishi I, Fujii S, Okamura H, Parmley T, Mori T. Development of interstitial cells and ovigerous cords in the human fetal ovary: an ultrastructural study. *J Anat*. 1986;148:121–135.

23. Pepling ME, Spradling AC. Female mouse germ cells form synchronously dividing cysts. *Development*. 1998;125(17):3323–3328.

24. Menke DB, Koubova J, Page DC. Sexual differentiation of germ cells in XX mouse gonads occurs in an anterior-to-posterior wave. *Dev Biol*. 2003;262(2):303–312.

25. Bullejos M, Koopman P. Germ cells enter meiosis in a rostro-caudal wave during development of the mouse ovary. *Mol Reprod Dev*. 2004;68(4):422–428.

26. Bowles J, Knight D, Smith C, et al. Retinoid signaling determines germ cell fate in mice. *Science*. 2006;312(5773):596–600.

27. Koubova J, Menke DB, Zhou Q, Capel B, Griswold MD, Page DC. Retinoic acid regulates sex-specific timing of meiotic initiation in mice. *Proc Natl Acad Sci USA*. 2006;103(8):2474–2479.

28. Anderson RA, Fulton N, Cowan G, Coutts S, Saunders PT. Conserved and divergent patterns of expression of DAZL, VASA and OCT4 in the germ cells of the human fetal ovary and testis.. *BMC Dev Biol*. 2007;7:136.

29. Wilhelm D, Yang JX, Thomas P. Mammalian sex determination and gonad development. *Curr Top Dev Biol*. 2013;106:89–121.

30. MacLean G, Li H, Metzger D, Chambon P, Petkovich M. Apoptotic extinction of germ cells in testes of Cyp26b1 knockout mice. *Endocrinology*. 2007;148(10):4560–4567.

31. Tingen C, Kim A, Woodruff TK. The primordial pool of follicles and nest breakdown in mammalian ovaries. *Mol Hum Reprod*. 2009;15(12):795–803.

32. Vanorny DA, Prasasya RD, Chalpe AJ, Kilen SM, Mayo KE. Notch signaling regulates ovarian follicle formation and coordinates follicular growth. *Mol Endocrinol.* 2014;28(4):499−511.

33. Sarraj MA, Drummond AE. Mammalian foetal ovarian development: consequences for health and disease. *Reproduction.* 2012;143(2):151−163.

34. Hirshfield AN. Heterogeneity of cell populations that contribute to the formation of primordial follicles in rats. *Biol Reprod.* 1992;47(3):466−472.

35. Mork L, Maatouk DM, McMahon JA, et al. Temporal differences in granulosa cell specification in the ovary reflect distinct follicle fates in mice. *Biol Reprod.* 2012;86(2):37.

36. van Wezel IL, Rodgers RJ. Morphological characterization of bovine primordial follicles and their environment *in vivo. Biol Reprod.* 1996;55(5):1003−1011.

37. Tingen CM, Bristol-Gould SK, Kiesewetter SE, Wellington JT, Shea L, Woodruff TK. Prepubertal primordial follicle loss in mice is not due to classical apoptotic pathways. *Biol Reprod.* 2009;81(1):16−25.

38. Kerr JB, Myers M, Anderson RA. The dynamics of the primordial follicle reserve. *Reproduction.* 2013;146(6):R205−R215.

39. Rodrigues P, Limback D, McGinnis LK, Plancha CE, Albertini DF. Multiple mechanisms of germ cell loss in the perinatal mouse ovary. *Reproduction.* 2009;137(4):709−720.

40. Baker TG. Radiosensitivity of mammalian oocytes with particular reference to the human female. *Am J Obstet Gynecol.* 1971;110(5):746−761.

41. Henderson SA, Edwards RG. Chiasma frequency and maternal age in mammals. *Nature.* 1968;218(5136):22−28.

42. Polani PE, Crolla JA. A test of the production line hypothesis of mammalian oogenesis. *Hum Genet.* 1991;88(1):64−70.

43. Tease C, Fisher G. Further examination of the production-line hypothesis in mouse foetal oocytes. 1. Inversion heterozygotes. *Chromosoma.* 1986;93(5):447−452.

44. Speed RM, Chandley AC. Meiosis in the foetal mouse ovary. II. Oocyte development and age-related aneuploidy. Does a production line exist?. *Chromosoma.* 1983;88(3):184−189.

45. Rowsey R, Gruhn J, Broman KW, Hunt PA, Hassold T. Examining variation in recombination levels in the human female: a test of the production-line hypothesis. *Am J Hum Genet.* 2014;95(1):108−112.

46. Fortune JE, Kito S, Wandji SA, Srsen V. Activation of bovine and baboon primordial follicles *in vitro. Theriogenology.* 1998;49(2):441−449.

47. Hsueh AJ, Kawamura K, Cheng Y, Fauser BC. Intraovarian control of early folliculogenesis. *Endocr Rev.* 2014;er20141020.

48. Reddy P, Shen L, Ren C, et al. Activation of Akt (PKB) and suppression of FKHRL1 in mouse and rat oocytes by stem cell factor during follicular activation and development. *Dev Biol.* 2005;281(2):160−170.

49. Reddy P, Zheng W, Liu K. Mechanisms maintaining the dormancy and survival of mammalian primordial follicles. *Trends Endocrinol Metab.* 2010;21(2):96−103.

50. Reddy P, Adhikari D, Zheng W, et al. PDK1 signaling in oocytes controls reproductive aging and lifespan by manipulating the survival of primordial follicles. *Hum Mol Genet.* 2009;18(15):2813−2824.

51. Adhikari D, Gorre N, Risal S, et al. The safe use of a PTEN inhibitor for the activation of dormant mouse primordial follicles and generation of fertilizable eggs. *PLoS One.* 2012;7 (6):e39034.

52. Li J, Kawamura K, Cheng Y, et al. Activation of dormant ovarian follicles to generate mature eggs. *Proc Natl Acad Sci USA*. 2010;107(22):10280–10284.

53. Kawamura K, Cheng Y, Suzuki N, et al. Hippo signaling disruption and Akt stimulation of ovarian follicles for infertility treatment. *Proc Natl Acad Sci USA*. 2013;110 (43):17474–17479.

54. McLaughlin M, Kinnell HL, Anderson RA, Telfer EE. Inhibition of phosphatase and tensin homologue (PTEN) in human ovary *in vitro* results in increased activation of primordial follicles but compromises development of growing follicles. *Mol Hum Reprod*. 2014;20(8):736–744.

55. Schmidt D, Ovitt CE, Anlag K, et al. The murine winged-helix transcription factor Foxl2 is required for granulosa cell differentiation and ovary maintenance. *Development*. 2004;131 (4):933–942.

56. Uda M, Ottolenghi C, Crisponi L, et al. Foxl2 disruption causes mouse ovarian failure by pervasive blockage of follicle development. *Hum Mol Genet*. 2004;13(11):1171–1181.

57. Rajkovic A, Pangas SA, Ballow D, Suzumori N, Matzuk MM. NOBOX deficiency disrupts early folliculogenesis and oocyte-specific gene expression. *Science*. 2004;305(5687):1157–1159.

58. Pangas SA, Choi Y, Ballow DJ, et al. Oogenesis requires germ cell-specific transcriptional regulators Sohlh1 and Lhx8. *Proc Natl Acad Sci USA*. 2006;103(21):8090–8095.

59. Choi Y, Yuan D, Rajkovic A. Germ cell-specific transcriptional regulator sohlh2 is essential for early mouse folliculogenesis and oocyte-specific gene expression. *Biol Reprod*. 2008;79(6):1176–1182.

60. Durlinger AL, Kramer P, Karels B, et al. Control of primordial follicle recruitment by anti-Müllerian hormone in the mouse ovary. *Endocrinology*. 1999;140(12):5789–5796.

61. Kevenaar ME, Meerasahib MF, Kramer P, et al. Serum anti-Müllerian hormone levels reflect the size of the primordial follicle pool in mice. *Endocrinology*. 2006;147(7):3228–3234.

62. Visser JA, de Jong FH, Laven JS, Themmen AP. Anti-Müllerian hormone: a new marker for ovarian function. *Reproduction*. 2006;131(1):1–9.

63. McDade TW, Woodruff TK, Huang YY, et al. Quantification of anti-Müllerian hormone (AMH) in dried blood spots: validation of a minimally invasive method for assessing ovarian reserve. *Hum Reprod*. 2012;27(8):2503–2508.

64. Baarends WM, Uilenbroek JT, Kramer P, et al. Anti-müllerian hormone and anti-müllerian hormone type II receptor messenger ribonucleic acid expression in rat ovaries during postnatal development, the estrous cycle, and gonadotropin-induced follicle growth. *Endocrinology*. 1995;136(11):4951–4962.

65. Gougeon A, Delangle A, Arouche N, Stridsberg M, Gotteland JP, Loumaye E. Kit ligand and the somatostatin receptor antagonist, BIM-23627, stimulate *in vitro* resting follicle growth in the neonatal mouse ovary. *Endocrinology*. 2010;151(3):1299–1309.

66. Holt JE, Jackson A, Roman SD, Aitken RJ, Koopman P, McLaughlin EA. CXCR4/SDF1 interaction inhibits the primordial to primary follicle transition in the neonatal mouse ovary. *Dev Biol*. 2006;293(2):449–460.

67. McGee EA, Hsu SY, Kaipia A, Hsueh AJ. Cell death and survival during ovarian follicle development. *Mol Cell Endocrinol*. 1998;140(1–2):15–18.

68. Zheng W, Zhang H, Gorre N, Risal S, Shen Y, Liu K. Two classes of ovarian primordial follicles exhibit distinct developmental dynamics and physiological functions. *Hum Mol Genet*. 2014;23(4):920–928.

69. Buccione R, Schroeder AC, Eppig JJ. Interactions between somatic cells and germ cells throughout mammalian oogenesis. *Biol Reprod*. 1990;43(4):543–547.

70. Eppig JJ, Wigglesworth K, Pendola FL. The mammalian oocyte orchestrates the rate of ovarian follicular development. *Proc Natl Acad Sci USA*. 2002;99(5):2890–2894.

71. Dong J, Albertini DF, Nishimori K, Kumar TR, Lu N, Matzuk MM. Growth differentiation factor-9 is required during early ovarian folliculogenesis. *Nature*. 1996;383(6600):531–535.

72. Yan C, Wang P, DeMayo J, et al. Synergistic roles of bone morphogenetic protein 15 and growth differentiation factor 9 in ovarian function. *Mol Endocrinol*. 2001;15(6):854–866.

73. McMahon HE, Hashimoto O, Mellon PL, Shimasaki S. Oocyte-specific overexpression of mouse bone morphogenetic protein-15 leads to accelerated folliculogenesis and an early onset of acyclicity in transgenic mice. *Endocrinology*. 2008;149(6):2807–2815.

74. Carabatsos MJ, Sellitto C, Goodenough DA, Albertini DF. Oocyte-granulosa cell heterologous gap junctions are required for the coordination of nuclear and cytoplasmic meiotic competence. *Dev Biol*. 2000;226(2):167–179.

75. Simon AM, Goodenough DA, Li E, Paul DL. Female infertility in mice lacking connexin 37. *Nature*. 1997;385(6616):525–529.

76. Ackert CL, Gittens JE, O'Brien MJ, Eppig JJ, Kidder GM. Intercellular communication via connexin43 gap junctions is required for ovarian folliculogenesis in the mouse. *Dev Biol*. 2001;233(2):258–270.

77. Sicinski P, Donaher JL, Geng Y, et al. Cyclin D2 is an FSH-responsive gene involved in gonadal cell proliferation and oncogenesis. *Nature*. 1996;384(6608):470–474.

78. Baker J, Hardy MP, Zhou J, et al. Effects of an Igf1 gene null mutation on mouse reproduction. *Mol Endocrinol*. 1996;10(7):903–918.

79. Freiman RN, Albright SR, Zheng S, Sha WC, Hammer RE, Tjian R. Requirement of tissue-selective TBP-associated factor TAFII105 in ovarian development. *Science*. 2001;293 (5537):2084–2087.

80. Voronina E, Lovasco LA, Gyuris A, Baumgartner RA, Parlow AF, Freiman RN. Ovarian granulosa cell survival and proliferation requires the gonad-selective TFIID subunit TAF4b. *Dev Biol*. 2007;303(2):715–726.

81. Falender AE, Shimada M, Lo YK, Richards JS. TAF4b, a TBP associated factor, is required for oocyte development and function. *Dev Biol*. 2005;288(2):405–419.

82. Hirshfield AN. Development of follicles in the mammalian ovary. *Int Rev Cytol*. 1991;124:43–101.

83. Hirshfield AN. Theca cells may be present at the outset of follicular growth. *Biol Reprod*. 1991;44(6):1157–1162.

84. Magoffin DA. Ovarian theca cell. *Int J Biochem Cell Biol*. 2005;37(7):1344–1349.

85. Elvin JA, Matzuk MM. Mouse models of ovarian failure. *Rev Reprod*. 1998;3(3):183–195.

86. Chun SY, Eisenhauer KM, Minami S, Billig H, Perlas E, Hsueh AJ. Hormonal regulation of apoptosis in early antral follicles: follicle-stimulating hormone as a major survival factor. *Endocrinology*. 1996;137(4):1447–1456.

87. Richards JS. Hormonal control of gene expression in the ovary. *Endocr Rev*. 1994;15 (6):725–751.

88. Rao MC, Midgley Jr. AR, Richards JS. Hormonal regulation of ovarian cellular proliferation. *Cell*. 1978;14(1):71–78.

89. Woodruff TK, D'Agostino J, Schwartz NB, Mayo KE. Dynamic changes in inhibin messenger RNAs in rat ovarian follicles during the reproductive cycle. *Science*. 1988;239 (4845):1296–1299.

90. Woodruff TK, D'Agostino J, Schwartz NB, Mayo KE. Decreased inhibin gene expression in preovulatory follicles requires primary gonadotropin surges. *Endocrinology*. 1989;124(5):2193−2199.

91. Stocco C, Telleria C, Gibori G. The molecular control of corpus luteum formation, function, and regression. *Endocr Rev*. 2007;28(1):117−149.

92. Devoto L, Fuentes A, Kohen P, et al. The human corpus luteum: life cycle and function in natural cycles. *Fertil Steril*. 2009;92(3):1067−1079.

93. Johnson J, Canning J, Kaneko T, Pru JK, Tilly JL. Germline stem cells and follicular renewal in the postnatal mammalian ovary. *Nature*. 2004;428(6979):145−150.

94. White YA, Woods DC, Takai Y, Ishihara O, Seki H, Tilly JL. Oocyte formation by mitotically active germ cells purified from ovaries of reproductive-age women. *Nat Med*. 2012;18(3):413−421.

95. Faddy MJ, Gosden RG. A model conforming the decline in follicle numbers to the age of menopause in women. *Hum Reprod*. 1996;11(7):1484−1486.

96. Hansen KR, Knowlton NS, Thyer AC, Charleston JS, Soules MR, Klein NA. A new model of reproductive aging: the decline in ovarian non-growing follicle number from birth to menopause. *Hum Reprod*. 2008;23(3):699−708.

97. Baker T.G., A Quantitative and Cytological Study of Germ Cells in Human Ovaries. Proceedings of the Royal Society of London Series B, Containing papers of a Biological character Royal Society. 1963;158:417−33.

98. Goswami D, Conway GS. Premature ovarian failure. *Hum Reprod Update*. 2005;11 (4):391−410.

99. Hanoux V, Pairault C, Bakalska M, Habert R, Livera G. Caspase-2 involvement during ionizing radiation-induced oocyte death in the mouse ovary. *Cell Death Differ*. 2007;14 (4):671−681.

100. Gonfloni S, Di Tella L, Caldarola S, et al. Inhibition of the c-Abl-TAp63 pathway protects mouse oocytes from chemotherapy-induced death. *Nat Med*. 2009;15(10):1179−1185.

101. Suh EK, Yang A, Kettenbach A, et al. p63 protects the female germ line during meiotic arrest. *Nature*. 2006;444(7119):624−628.

102. Ashwood-Smith MJ, Edwards RG. DNA repair by oocytes. *Mol Hum Reprod*. 1996;2(1):46−51.

103. Tilly JL. Commuting the death sentence: how oocytes strive to survive. *Nat Rev Mol Cell Biol*. 2001;2(11):838−848.

104. Meirow D, Nugent D. The effects of radiotherapy and chemotherapy on female reproduction. *Hum Reprod Update*. 2001;7(6):535−543.

105. Jeruss JS, Woodruff TK. Preservation of fertility in patients with cancer. *N Engl J Med*. 2009;360(9):902−911.

106. Kerr JB, Hutt KJ, Michalak EM, et al. DNA damage-induced primordial follicle oocyte apoptosis and loss of fertility require TAp63-mediated induction of Puma and Noxa. *Mol Cell*. 2012;48(3):343−352.

107. Kaipia A, Hsueh AJ. Regulation of ovarian follicle atresia. *Annu Rev Physiol*. 1997;59:349−363.

108. De Pol A, Vaccina F, Forabosco A, Cavazzuti E, Marzona L. Apoptosis of germ cells during human prenatal oogenesis. *Hum Reprod*. 1997;12(10):2235−2241.

109. Coucouvanis EC, Sherwood SW, Carswell-Crumpton C, Spack EG, Jones PP. Evidence that the mechanism of prenatal germ cell death in the mouse is apoptosis. *Exp Cell Res*. 1993;209(2):238−247.

110. Xu B, Hua J, Zhang Y, et al. Proliferating cell nuclear antigen (PCNA) regulates primordial follicle assembly by promoting apoptosis of oocytes in fetal and neonatal mouse ovaries. *PLoS One*. 2011;6(1):e16046.

111. De Felici M, Lobascio AM, Klinger FG. Cell death in fetal oocytes: many players for multiple pathways. *Autophagy*. 2008;4(2):240–242.

112. Ohno S, Smith JB. Role of fetal follicular cells in meiosis of mammalian ooecytes. *Cytogenetics*. 1964;3:324–333.

113. Ratts VS, Flaws JA, Kolp R, Sorenson CM, Tilly JL. Ablation of bcl-2 gene expression decreases the numbers of oocytes and primordial follicles established in the postnatal female mouse gonad. *Endocrinology*. 1995;136(8):3665–3668.

114. Greenfeld CR, Pepling ME, Babus JK, Furth PA, Flaws JA. BAX regulates follicular endowment in mice. *Reproduction*. 2007;133(5):865–876.

115. Perez GI, Jurisicova A, Wise L, et al. Absence of the proapoptotic Bax protein extends fertility and alleviates age-related health complications in female mice. *Proc Natl Acad Sci USA*. 2007;104(12):5229–5234.

116. Perez GI, Robles R, Knudson CM, Flaws JA, Korsmeyer SJ, Tilly JL. Prolongation of ovarian lifespan into advanced chronological age by Bax-deficiency. *Nat Genet*. 1999;21(2):200–203.

117. Bergeron L, Perez GI, Macdonald G, et al. Defects in regulation of apoptosis in caspase-2-deficient mice. *Genes Dev*. 1998;12(9):1304–1314.

118. Kim SY, Cordeiro MH, Serna VA, et al. Rescue of platinum-damaged oocytes from programmed cell death through inactivation of the p53 family signaling network. *Cell Death Differ*. 2013;20(8):987–997.

119. Livera G, Petre-Lazar B, Guerquin MJ, Trautmann E, Coffigny H, Habert R. p63 null mutation protects mouse oocytes from radio-induced apoptosis. *Reproduction*. 2008;135(1):3–12.

120. Luoh SW, Bain PA, Polakiewicz RD, et al. Ztx mutation results in small animal size and reduced germ cell number in male and female mice. *Development*. 1997;124(11):2275–2284.

121. Ahn RW, Barrett SL, Raja MR, et al. Nano-encapsulation of arsenic trioxide enhances efficacy against murine lymphoma model while minimizing its impact on ovarian reserve *in vitro* and *in vivo*. *PLoS One*. 2013;8(3):e58491.

CHAPTER 2

Relevant Cancer Diagnoses, Commonly Used Chemotherapy Agents and Their Biochemical Mechanisms of Action

Christopher H. Grant and Charlie Gourley
Institute of Genetics and Molecular Medicine, University of Edinburgh Cancer Research UK Centre, Western General Hospital, Edinburgh, UK

2.1 INTRODUCTION

More than 75% of children treated for malignant disease will be alive 10 years following initial diagnosis in the UK.[1] In addition, many women of reproductive age will undergo curative treatment for malignant disease. An understanding of the impact that these treatments will have on their future reproductive health is critical in order to inform both treatment choices and counselling with respect to educational, employment and family-planning decisions.

The number of primordial follicles in the human ovary is highest at 5 months gestation, following which the rate of loss progressively accelerates, particularly in the 10 years prior to the menopause (Figure 2.1).[2] Radiotherapy and chemotherapy both have an impact on follicle numbers and, as such, can affect future female fertility. This includes acute ovarian failure, or premature ovarian insufficiency (POI) (either loss of menses as in acute ovarian failure or menstruation within the context of a damaged ovary). Some studies have focused on the incidence of acute ovarian failure as a marker of impact on fertility, but other mechanisms, particularly later POI, significantly affect female fertility following cancer treatment.[3] In healthy women, fertility is lost one decade before natural menopause, with a decline over the preceding decade.[4] Therefore, in cancer patients who experience POI, it is likely that their fertility decreases markedly several years before this. As such, the best marker of the reproductive capacity following cancer treatment is fertility itself rather than incidence of POI.[5]

Cancer Treatment and the Ovary. DOI: http://dx.doi.org/10.1016/B978-0-12-801591-9.00002-3

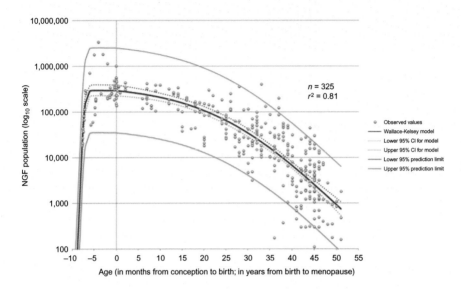

Figure 2.1 **Model of ovarian reserve that best fits histological data according to Wallace and Kelsey.** *The Y axis illustrates the number of primordial follicles constituting the ovarian reserve. There is large interpatient variation in ovarian reserve throughout the reproductive years. NGF, Non-growing follicles. This figure is reproduced from[2], published under an open-access license by PLoS.*

Because a range of terms are used in the literature for describing ovarian damage, the terminology used in the original report will be used here.

The cancers whose treatment most often results in female infertility are leukaemia, Hodgkin lymphoma, non-Hodgkin lymphoma, breast carcinoma, sarcomas and germ cell tumours. The incidence of female infertility within each cancer type depends upon the age at diagnosis, whether surgery or radiotherapy affects the reproductive organs (including the hypothalamic–pituitary axis) and the nature of the chemotherapy that is used. The combination of chemotherapy and radiotherapy is more likely to have a significant impact on fertility but, for a given individual, precisely predicting this impact is difficult, particularly given the large inter-patient variation in ovarian reserve (discussed further in Chapter 3). Over the course of the last decade there has been increasing use of targeted biological therapies to treat cancer. There has been less experience with these and the long-term consequences of their exposure on female fertility are unclear. There is some suggestion that bevacizumab (one of the more common biological agents, used in the treatment of colorectal, lung, breast, ovarian, cervical, brain and renal cancer) can cause permanent ovarian failure.

A retrospective questionnaire-based survey of 5-year survivors from the Childhood Cancer Survivor Study demonstrated that they had a decreased incidence of becoming pregnant compared to female siblings (OR, 0.81; 95% CI, 0.73−0.90, p < 0.001).[6] In this study, acute ovarian failure occurred in 6.3% of eligible survivors.[7] High-dose radiation (>10 Gy) and exposure to alkylating agents were all significant risk factors for acute ovarian failure. In addition, premature non-surgical menopause occurred in 8% of patients compared to 0.8% of siblings. Risk factors for premature menopause include: attained age, exposure to increased dose of radiation, exposure to alkylating agents and high-dose chemotherapy. The British Childhood Cancer Survivor study retrospectively assessed the fertility of 5133 females by questionnaire.[8] The number of live births was two-thirds of that expected (O/E, 0.64; 95% CI, 0.62−0.66) and was lowest amongst survivors treated with brain (O/E, 0.52; 95% CI, 0.48−0.56) and abdominal irradiation (O/E, 0.55; 95% CI, 0.50−0.61).

Here we describe the impact of radiotherapy and chemotherapy on future female fertility and then summarize the mechanisms of action of different classes of chemotherapy drugs in order to form the context for the discussion of the specific effects of these agents on the ovary in subsequent chapters.

2.2 IMPACT OF RADIOTHERAPY ON FUTURE FEMALE FERTILITY

In female premenopausal cancer patients, any radiotherapy plan that affects the ovaries (whether that is as part of pelvic, abdominal or whole-body irradiation) will have an impact on fertility. The extent of this impact is dependent on patient age, dose, radiotherapy field and fractionation schedule. Radiotherapy-induced primordial follicle oocyte loss depends upon the size of the oocyte pool. As such, the younger the patient is at the time of radiotherapy the later the onset of POI (for a given dose of irradiation).[9] Irradiation of the uterus can result in decreased fertility probably because of effects on the uterine musculature and vasculature, although the exact mechanisms and the extent of this impact (as discrete from the effects of ovarian irradiation) are poorly defined. Cranial irradiation can also have a profound impact on fertility through its effect on circulating gonadotropin levels. However, use of exogenous gondotropins or gonadotropin-releasing hormone can be used to treat this cause of radiation-induced infertility.

Multivariate analysis of the female fertility rate in the Childhood Cancer Survivor Study showed that it was significantly decreased in women previously exposed to a hypothalamic/pituitary radiotherapy dose ≥30 Gy (RR, 0.61; 95% CI, 0.44−0.83) or an ovarian/uterine dose of >5 Gy (RR, 0.56 for 5−10 Gy; 95% CI, 0.37−0.85; and RR, 0.18 for >10 Gy; 95% CI, 0.13−0.26).[6] In a more specific analysis of the same patient group, multivariable analysis revealed that female cancer survivors had a decreased chance of pregnancy with hypothalamic/pituitary doses >22 Gy compared to those with no hypothalamic/pituitary radiotherapy. There is no indication regarding the extent to which this is related to the direct impact of hypothalamic/pituitary radiation and the extent to which this is related to other factors, e.g., impact on growth, progression through puberty, etc.[10] The potential for selection bias is a factor that must be borne in mind when interpreting the results of these questionnaire-based studies.

2.3 IMPACT OF CHEMOTHERAPY ON FUTURE FEMALE FERTILITY

In a retrospective survey-based analysis of reproductive-age Californian women treated with chemotherapy for cancer, Letourneau et al. found the incidence of acute ovarian failure to be 10% in non-Hodgkin lymphoma, 9% in breast cancer, 8% in Hodgkin lymphoma, 5% in gastrointestinal cancers and 3% in leukaemia.[5] The proportion of women experiencing acute ovarian failure significantly increased with older age at diagnosis. For patients whose menses returned within 12 months of treatment, there was an association between early menopause and younger age at diagnosis.

Multivariate analysis of the female fertility rate in the Childhood Cancer Survivor Study showed that patients who received higher doses of alkylating agents were less likely to become pregnant compared to selected female siblings.[6]

A largely retrospective study in Asian non-metastatic breast cancer patients showed that patient age at chemotherapy treatment was the only statistically significant factor in determining the incidence of chemotherapy-induced ovarian failure and reversible amenorrhoea.[11]

In this study, the overall incidence of amenorrhoea was high, both at the end of chemotherapy (93%) and 12 months following chemotherapy.

A retrospective analysis of a large cohort of Hodgkin lymphoma survivors treated in the European Organisation for Research and Treatment of Cancer (EORTC) and Group d'Etude des Lymphomes de l'Adulte trials between 1964 and 2004 identified premature ovarian failure in 60% of patients (95% CI, 41−79%).[12] The incidence was, however, only 3% (95% CI, 1−7%) after non-alkylating agent chemotherapy regimes. There was a linear relationship between premature ovarian failure and both use of alkylating agent chemotherapy and age at treatment. In patients receiving non-alkylating chemotherapy, while the incidence of premature ovarian failure was only 3% if treated before the age of 32 years, it was 9% if treated after the age of 32. If menstruation returned after treatment then the premature ovarian failure risk was independent of age at treatment in this study (in contrast to the study by Letourneau *et al.* described above). In terms of the impact on subsequent fertility amongst women who ultimately developed premature ovarian failure, 22% had a child after treatment, compared to 41% who did not later develop premature ovarian failure.

There is some evidence that non-alkylating chemotherapy carries only a small additional risk of POI. Patients treated with pelvic radiotherapy and/or alkylating agents are clearly at increased risk. It is important in the counselling of these women to emphasize the impact on family planning, because although menstruation may return after chemotherapy the possibility of subsequent early menopause needs to be accounted for. This window of opportunity for fertility may also have implications for educational and career choices.

2.4 MECHANISMS OF ACTION OF THE COMMONLY USED CHEMOTHERAPY DRUGS

Chemotherapy is the common term used to describe cytotoxic drugs, which generally act by selectively damaging or killing rapidly dividing cells. There are around 30−40 chemotherapeutic agents commonly employed in routine oncological practice (their development is well reviewed by Chabner and Roberts).[13] Tumour cells are usually rapidly dividing and so these agents act to reduce the rate of tumour growth and/or reduce tumour size. A large number of normal host cells, including those of the ovary, are also in a state of cell division and these cells are also affected by cytotoxic therapy. Chemotherapy drugs can either be given as single agents or in combination. The rationale for the use of combination chemotherapy is that some agents have synergistic effects increasing efficacy, and, theoretically, agents with different mechanisms of action reduce the likelihood of cancer cell resistance developing.[14,15] Of course, the use of agents that have some overlap of toxicity profiles increases the impact on non-tumour cells, including those of the reproductive system. Chemotherapy is usually administered in cycles whereby following a pulse of therapy, there is a period of days or a week before a subsequent pulse is administered. This is in order to allow essential normal cells (such as those of the bone marrow and the gastrointestinal tract) to recover, with the assumption that recovery in tumour cells is less efficient.[13] These regimes are designed with anti-neoplastic effect as the priority, and generally revolve around the principle that as much chemotherapy as possible should be given within a set period of time without exposing the patients to an unacceptable risk in terms of other toxicities. In cancers where there is a clear relationship between dose and chance of cure, a myeloablative dose of chemotherapy can be given supported by a stem cell or bone marrow transplant. To date, little consideration has been given to the impact on female reproductive function in order to achieve these oncological goals.

The mechanisms of action of the major categories of chemotherapy drugs are described below. Examples of agents that fall into each category are shown in Table 2.1.

2.4.1 Alkylating Agents

These highly reactive compounds form irreversible covalent bonds with amino, carboxyl, sulfhydryl and phosphate groups in DNA, RNA and proteins.[15] The number and location of bonding sites are drug specific.[16] Interstrand and intrastrand crosslinking of DNA has the effect of preventing replication. In animal studies, these agents have been shown to have an effect on mitochondria, reducing the transmembrane potential, causing cytochrome c release into the cytoplasm and hence apoptosis.[17] Alkylating agents do not act on a specific phase of the cell cycle, but rather are dependent on ongoing cellular proliferation for therapeutic effect.[13,15]

Table 2.1 List of Chemotherapy Agents by Category (Columns) and Classes within Each Category

Alkylating Agents	Antimetabolites	Antitumour Antibiotics	Mitotic Inhibitors	Platinum Drugs	Topoisomerase Inhibitors
ALKYL SULFONATES	Capecitabine	ANTHRACYCLINES*	EPOTHILONES	Carboplatin	TOPO 1 INHIBITORS
Busulfan	Cladribine	Daunorubicin	Ixabepilone	Cisplatin	Irinotecan
ETHYLENIMINES	Clofarabine	Doxorubicin	TAXANES	Oxaliplatin	Topotecan
Altretamine	Cytarabine	Epirubicin	Docetaxel		TOPO 2 INHIBITORS
Thiotepa	Floxuridine	Idarubicin	Paclitaxel		Etoposide
NITROGEN MUSTARDS	Fludarabine	Mitoxantrone	VINCA ALKALOIDS		Teniposide
Chlorambucil	5-Fluorouracil	OTHERS	Vinblastine		
Cyclophosphamide	Gemcitabine	Actinomycin D	Vincristine		
Ifosfamide	Hydroxyurea	Bleomycin	Vinorelbine		
Mechlorethamine	6-Mercaptopurine	Mitomycin C	OTHERS		
Melphalan	Methotrexate	Trabectedin	Estramustine		
NITROSOUREAS	Pemetrexed				
Carmustine (BCNU)	Pentostatin				
Lomustine	Thioguanine				
Streptozocin					
HYDRAZINES/TRIAZINES					
Dacarbazine (DTIC)					
Procarbazine					
Temozolomide					

*Many of the anthracyclines also inhibit topoisomerase 1 and 1 or 2.

The main classes of alkylating agents are: alkyl sulfonates (e.g., busulfan); ethylenamines (e.g., thiotepa); nitrogen mustards (e.g., cyclophosphamide, ifosfamide); nitrosoureas (e.g., carmustine) and triazines (e.g., temozolamide). These drugs are used to treat leukaemia, lymphoma, sarcoma, breast, ovarian and brain cancer.

There is considerable retrospective data that suggest alkylating agents are the most ovotoxic of all chemotherapy. POI occurs in 60–80% of breast cancer patients treated with cyclophosphamide, methotrexate and 5-fluorouracil,[18,19] with that toxicity largely being attributed to the cyclophosphamide component.[20] There is clear evidence of an association between dose and early menopause,[21] which is of particular relevance when alkylating agents are used in high-dose chemotherapy regimes supported by bone marrow or stem cell transplantation (see below). These agents (particularly cyclophosphamide) also affect mitochondria and hence it is postulated that more metabolically active cells are preferentially affected (e.g., granulosa cells in the ovary).[22]

2.4.2 Antimetabolites

These are structural analogues of the normal components of anabolism of nucleic acid synthesis. Hence, they are taken up for integration into the normal DNA and RNA synthetic process of the cancer cell, but they then inhibit its normal function.[23] These drugs are most active in the S phase of the cell cycle and exhibit a dose-independent response.[15] They can be split into three categories: antifolates, pyrimidine analogues and purine analogues. Methotrexate (an antifolate) is converted to methotrexate polyglutamate in the cell, and binds to dihydrofolate reductase, which in turn inhibits thymidylate purine biosynthesis and results in induction of apoptosis. The nucleoside analogues (5-fluorouracil, gemcitabine, cytosine arabinoside, 6-mercaptopurine) undergo three steps of phosphorylation on entering the cell. The resulting nucleoside triphosphate can then be incorporated into DNA/RNA or alternatively can inhibit enzymes involved in nucleic acid synthesis, such as DNA polymerases or ribonucleotide reductases.[24] The resultant inhibition of DNA synthesis causes apoptosis. This mechanism of action would suggest that the most likely cells within the ovary to be affected by these agents would be granulosa cells, but there is a lack of direct evidence demonstrating this. Antimetabolites are used to treat cancers including leukaemia, lymphoma, sarcoma, breast, gastrointestinal, lung, bladder and ovarian cancer.

2.4.3 Antitumour Antibiotics

Anthracyclines are antibiotics that are derived from a species of fungus and have multi-modal mechanisms of action. They intercalate in DNA, forming free radical intermediates,[25] damaging DNA and disturbing its synthesis. They are also topoisomerase inhibitors (particularly topoisomerase 2).[14,15] Some studies suggest that anthracyclines have an affinity for cardiolipin, which is expressed on the inner mitochondrial membrane, thereby bringing about disruption of the electron transport chain and release of cytochrome c into the cytoplasm, which in turn leads to apoptosis.[26] They function in all phases of the cell cycle. They are used to treat a number of cancers including lymphomas, breast cancer and sarcomas.

Other antitumour antibiotics include: bleomycin, which binds to DNA, leading to DNA fragmentation and may also lead to the formation of oxygen free radicals[16]; and actinomycin D, which binds to DNA, resulting in inhibition of transcription.[14] Trabectedin, or ET-743, originally derived from the marine tunicate *Ecteinascidia turbinate*, binds to the minor groove of DNA and disrupts both transcription and DNA repair.[27,28] It causes G2 arrest and a delay in progression through S phase of the cell cycle.[29] These agents are used for treating germ cell tumours, sarcomas and haematological malignancies.

Anthracyclines have been largely considered only mildly ovotoxic, but one recent study[5] suggested that previous assessments may have underestimated the effect. Given the mechanisms of action described above, mitotically and metabolically active cells are likely to be more affected and therefore, in terms of ovotoxicity, this may be more likely to be due to effects on granulosa cells than on oocytes.

2.4.4 Mitotic Inhibitors

These agents are often derived from plant alkaloids or other natural products. They act in various fashions in order to prevent mitosis (i.e., during M phase of the cell cycle). Vinca alkaloids (e.g., vincristine, vinblastine) inhibit the formation of the mitotic spindle that is required to pull replicated chromosomes to opposite poles of a dividing cell during mitosis. Taxanes (e.g., docetaxel, paclitaxel) bind to and stabilize the microtubule network, thereby preventing its normal reorganization and dissociation. Again, given this mechanism of action, it is likely that ovotoxicity results from the effect upon mitotically active granulosa cells rather than the effect upon oocytes. These agents are used to treat a wide range of cancers including leukaemias, lymphomas, germ cell tumours, sarcomas, breast, lung and ovarian cancers.

2.4.5 Platinum Drugs
Platinum compounds (e.g., carboplatin, cisplatin, oxaliplatin) have been central to the practice of oncology for the last 40 years.[30] Cisplatin and carboplatin covalently bind to DNA causing intrastrand and interstrand DNA adducts. This results in restriction of DNA replication, transcription, cell cycle arrest and programmed cell death.[31] While the primary mechanism of action of cisplatin and carboplatin is mediated through the induction of DNA damage, there also appears to be an effect on the intrinsic mitochondrial pathway and a component of endoplasmic reticulum stress which can both result in apoptosis.[32] Oxaliplatin has a similar mechanism of action to cisplatin, but there are some differences in the crosslinks formed[33] and, perhaps as a result of this, it has some activity in cisplatin-resistant cells.[34]

Platinum agents are used to treat lymphomas, sarcomas, breast, colorectal, germ cell, ovarian, lung, gastro-oesophageal and bladder cancers. Most retrospective data suggests that the impact of cisplatin treatment on female fertility is very much lower than alkylating agents, with some studies suggesting that there is no increased risk of premature ovarian failure.[20]

2.4.6 Topoisomerase Inhibitors
These agents interfere with topoisomerase 1 and 2, enzymes that are responsible for the cleavage and physical stability of DNA strands.[35] Topoisomerase 1 inhibitors such as topotecan and irinotecan bind to the enzyme–DNA complex preventing DNA unwinding, replication and transcription. Topoisomerase 2 inhibitors such as etoposide bind to and stabilize the topoisomerase-2–DNA complex. This results in DNA strand breaks and blocks on DNA replication with resultant apoptosis.[13–15] These agents principally affect cells during the S phase of the cell cycle. There is limited data regarding the ovarian cell type that is predominantly damaged by exposure to topoisomerase inhibitors. These drugs are used to treat lymphomas, sarcomas, colorectal, lung, breast and germ cell tumours.

2.4.7 High-Dose Chemotherapy
In some malignancies and for some agents there is a clear association between dose of treatment and cancer cell kill. In some diseases

(e.g., leukaemia, lymphoma, germ cell tumours) the ability to deliver a dose that will ablate the bone marrow can determine whether a patient will be cured or not. These treatments are supported by stem cell or bone marrow transplants following the chemotherapy in order that a functional immune system can be restored. Alkylating agents often form a significant component of these regimes and as such the impact on future fertility can be very significant.

2.4.8 Targeted Therapies

As the molecular abnormalities underlying many of the common malignancies are being uncovered, targeted molecular therapies are increasingly being tested or used in routine practice across most cancer types. These agents block oncogene-activated signal transduction pathways, pathways involved in angiogenesis or DNA-repair pathways that cancers have become particularly "dependent" upon (a principle known as "oncogene addiction"). Although these drugs are becoming more mainstream, the longest-established agents have been used for no more than 15 years (in most cases the experience is much shorter than that). As such, the impact of these agents on female fertility remains uncertain. However, there is a suggestion that bevacizumab, a recombinant humanized monoclonal antibody that targets vascular endothelial growth factor (VEGF)-A, may have an impact on female fertility. This agent is used for the treatment of colorectal, lung, breast, ovarian, cervical, brain and renal cancer. The incidence of ovarian failure was found to be higher (34% vs. 2%) in premenopausal women receiving bevacizumab in combination with mFolfox chemotherapy (folinic acid, 5-fluorouracil and oxaliplatin) compared to those receiving mFolfox chemotherapy alone for adjuvant treatment for colorectal cancer.[36] Validation of this finding is required. Although the underlying mechanism remains unknown, blood vessel damage resulting in local ischaemia and cortical fibrosis is a possibility.[37,38] Further post-marketing studies will be required in order to establish the impact of these novel agents on female infertility.

REFERENCES

1. Cancerstats, CRU. Childhood cancer survival statistics, 2015.

2. Wallace WH, Kelsey TW. Human ovarian reserve from conception to the menopause. *PLoS One.* 2010;5(1):e8772.

3. Sonmezer M, Oktay K. Fertility preservation in female patients. *Hum Reprod Update.* 2004;10 (3):251−266.

4. te Velde ER, Pearson PL. The variability of female reproductive ageing. *Hum Reprod Update.* 2002;8(2):141–154.

5. Letourneau JM, Ebbel EE, Katz PP, et al. Acute ovarian failure underestimates age-specific reproductive impairment for young women undergoing chemotherapy for cancer. *Cancer.* 2012;118(7):1933–1939.

6. Green DM, Kawashima T, Stovall M, et al. Fertility of female survivors of childhood cancer: a report from the childhood cancer survivor study. *J Clin Oncol.* 2009;27(16):2677–2685.

7. Green DM, Sklar CA, Boice Jr. JD, et al. Ovarian failure and reproductive outcomes after childhood cancer treatment: results from the Childhood Cancer Survivor Study. *J Clin Oncol.* 2009;27(14):2374–2381.

8. Reulen RC, Zeegers MP, Wallace WH, et al. British Childhood Cancer Survivor, S. Pregnancy outcomes among adult survivors of childhood cancer in the British Childhood Cancer Survivor Study. *Cancer Epidemiol Biomarkers Prev.* 2009;18(8):2239–2247.

9. Wallace WH, Thomson AB, Kelsey TW. The radiosensitivity of the human oocyte. *Hum Reprod.* 2003;18(1):117–121.

10. Green DM, Nolan VG, Kawashima T, et al. Decreased fertility among female childhood cancer survivors who received 22–27 Gy hypothalamic/pituitary irradiation: a report from the Childhood Cancer Survivor Study. *Fertil Steril.* 2011;95(6):1922–1927:7, e1.

11. Tiong V, Rozita AM, Taib NA, Yip CH, Ng CH. Incidence of chemotherapy-induced ovarian failure in premenopausal women undergoing chemotherapy for breast cancer. *World J Surg.* 2014;38(9):2288–2296.

12. van der Kaaij MA, Heutte N, Meijnders P, et al. Premature ovarian failure and fertility in long-term survivors of Hodgkin's lymphoma: a European Organisation for Research and Treatment of Cancer Lymphoma Group and Groupe d'Etude des Lymphomes de l'Adulte Cohort Study. *J Clin Oncol.* 2012;30(3):291–299.

13. Chabner BA, Roberts Jr. TG. Timeline: chemotherapy and the war on cancer. *Nat Rev Cancer.* 2005;5(1):65–72.

14. Gourley C. Cancer therapeutics. In: McKay GA, Reid JL, Walters MR, eds. *In Lecture Notes: Clinical Pharmacology and Therapeutics 8 edit.* Chichester: John Wiley and Sons Limited; 2010:205–216.

15. Malhotra V, Perry MC. Classical chemotherapy: mechanisms, toxicities and the therapeutic window. *Cancer Biol Ther.* 2003;2(4 suppl 1):S2–S4.

16. Payne S, Miles D. Mechanisms of anticancer drugs. In: Gleeson M, ed. *In Scott-Brown's Otorhinolaryngology: Head and Neck Surgery.* 7th ed. London: Hodder Arnold; 2008:34–46.

17. Zhao XJ, Huang YH, Yu YC, Xin XY. GnRH antagonist cetrorelix inhibits mitochondria-dependent apoptosis triggered by chemotherapy in granulosa cells of rats. *Gynecol Oncol.* 2010;118(1):69–75.

18. Bines J, Oleske DM, Cobleigh MA. Ovarian function in premenopausal women treated with adjuvant chemotherapy for breast cancer. *J Clin Oncol.* 1996;14(5):1718–1729.

19. Lower EE, Blau R, Gazder P, Tummala R. The risk of premature menopause induced by chemotherapy for early breast cancer. *J Womens Health Gend Based Med.* 1999;8(7):949–954.

20. Meirow D. Reproduction post-chemotherapy in young cancer patients. *Mol Cell Endocrinol.* 2000;169(1–2):123–131.

21. Chiarelli AM, Marrett LD, Darlington G. Early menopause and infertility in females after treatment for childhood cancer diagnosed in 1964–1988 in Ontario, Canada. *Am J Epidemiol.* 1999;150(3):245–254.

22. Morgan S, Anderson RA, Gourley C, Wallace WH, Spears N. How do chemotherapeutic agents damage the ovary? *Hum Reprod Update.* 2012;18(5):525–535.

23. Espinosa E, Zamora P, Feliu J, Gonzalez Baron M. Classification of anticancer drugs–a new system based on therapeutic targets. *Cancer Treat Rev.* 2003;29(6):515–523.

24. Galmarini CM, Mackey JR, Dumontet C. Nucleoside analogues and nucleobases in cancer treatment. *Lancet Oncol.* 2002;3(7):415–424.

25. Tan X, Wang DB, Lu X, et al. Doxorubicin induces apoptosis in H9c2 cardiomyocytes: role of overexpressed eukaryotic translation initiation factor 5A. *Biol Pharm Bull.* 2010; 33(10):1666–1672.

26. Pointon AV, Walker TM, Phillips KM, et al. Doxorubicin in vivo rapidly alters expression and translation of myocardial electron transport chain genes, leads to ATP loss and caspase 3 activation. *PLoS One.* 2010;5(9):e12733.

27. Minuzzo M, Marchini S, Broggini M, Faircloth G, D'Incalci M, Mantovani R. Interference of transcriptional activation by the antineoplastic drug ecteinascidin-743. *Proc Natl Acad Sci USA.* 2000;97(12):6780–6784.

28. Molinski TF, Dalisay DS, Lievens SL, Saludes JP. Drug development from marine natural products. *Nat Rev Drug Discov.* 2009;8(1):69–85.

29. Erba E, Bergamaschi D, Bassano L, et al. Ecteinascidin-743 (ET-743), a natural marine compound, with a unique mechanism of action. *Eur J Cancer.* 2001;37(1):97–105.

30. Kelland L. The resurgence of platinum-based cancer chemotherapy. *Nat Rev Cancer.* 2007;7(8):573–584.

31. Siddik ZH. Cisplatin: mode of cytotoxic action and molecular basis of resistance. *Oncogene.* 2003;22(47):7265–7279.

32. Mandic A, Hansson J, Linder S, Shoshan MC. Cisplatin induces endoplasmic reticulum stress and nucleus-independent apoptotic signaling. *J Biol Chem.* 2003;278(11):9100–9106.

33. Spingler B, Whittington DA, Lippard SJ. 2.4 A crystal structure of an oxaliplatin 1,2-d (GpG) intrastrand cross-link in a DNA dodecamer duplex. *Inorg Chem.* 2001;40 (22):5596–5602.

34. Raymond E, Faivre S, Chaney S, Woynarowski J, Cvitkovic E. Cellular and molecular pharmacology of oxaliplatin. *Mol Cancer Ther.* 2002;1(3):227–235.

35. Nussbaumer S, Bonnabry P, Veuthey JL, Fleury-Souverain S. Analysis of anticancer drugs: a review. *Talanta.* 2011;85(5):2265–2289.

36. Administration, USFaD. Avastin (bevacizumab). Detailed View: Safety Labeling Changes Approved By FDA Center for Drug Evaluation and Research (CDER). In MedWatch The FDA Safety Information and Adverse Event Reporting Program, Silver Spring, 2013.

37. Meirow D, Dor J, Kaufman B, et al. Cortical fibrosis and blood-vessels damage in human ovaries exposed to chemotherapy. Potential mechanisms of ovarian injury. *Hum Reprod.* 2007;22(6):1626–1633.

38. Nicosia SV, Matus-Ridley M, Meadows AT. Gonadal effects of cancer therapy in girls. *Cancer.* 1985;55(10):2364–2372.

CHAPTER 3

Clinical Assessment of Ovarian Toxicity

Richard A. Anderson
MRC Centre for Reproductive Health, University of Edinburgh, Edinburgh, UK

3.1 INTRODUCTION

Assessment of ovarian toxicity in the clinical setting is important in determining the effects of exposures on individual women and to compare different exposures in groups of women, for example those on different chemotherapy regimens. Such knowledge can also be used prospectively to inform discussion about likely effects on future fertility, for example in patients facing cancer therapies or other treatments involving cytotoxic agents, e.g., in rheumatic conditions such as systemic lupus erythematosus. This approach is also potentially applicable to prepubertal girls; up until now, however, this has proved very difficult in the absence of being able to assess the prepubertal ovary, although that is now changing as will be described below.

The essential basis of assessment is that the human ovary contains a finite pool of primordial follicles, which are formed *in utero*. A small number of follicles is activated daily into the growth phase, proceeding through preantral and subsequently the antral stages of development, becoming increasing dependent on follicle-stimulating hormone (FSH) and luteinizing hormone (LH) from the pituitary. While most follicles are destined to become atretic, increasing selection results in a single follicle being selected to become dominant each menstrual cycle to become the ovulatory follicle.[1] While numerically the smallest stages of follicle are by far the most abundant in the ovary, it is at the later antral and preovulatory stages that the follicle is most endocrinologically active, which is necessary for preparation for potential pregnancy. While a constant supply of small preantral follicles is consequently required to support the selection process that ultimately leads to ovulation, the number of small follicles declines throughout a woman's reproductive life,[2] whereas she continues to produce a single ovulatory

Cancer Treatment and the Ovary. DOI: http://dx.doi.org/10.1016/B978-0-12-801591-9.00003-5

follicle until the menopause. This underpins the absence of a clear relationship between ovulatory activity, i.e., oestrogen production, and the ovarian reserve.

3.2 THE USE OF CHEMOTHERAPY-RELATED AMENORRHOEA

The presence of cyclic menstruation, itself reflecting oestrogen production by the large antral and preovulatory follicle population, is the most readily apparent index of ovarian activity, with its absence indicating the loss of ovarian function. Chemotherapy-related amenorrhoea (CRA) is the most widely used index of ovarian toxicity in cancer studies and it undoubtedly remains a useful measure of assessment.[3] It is generally used to reflect premature ovarian insufficiency (POI), previously termed "premature menopause" or "premature ovarian failure". It is, therefore, taken to indicate the absence of follicular growth in the ovary and, by implication, depletion of the primordial pool, i.e., the ovarian reserve. While it should be noted that amenorrhoea can also result from a lack of gonadotropins, which may occur after, for example, cranial irradiation, in the great majority of instances the toxicity will clearly be to the ovary and the interpretation that it does indeed indicate POI is thus fairly robust. The precise definition does, however, vary significantly between studies (e.g., how many months of amenorrhoea are required). It is also important to recognize in this context that the amenorrhoea results from lack of oestrogen stimulation of the endometrium. As growing (oestrogen-producing) follicles are generally sensitive to chemotherapy, the acute toxicity to these follicles can result in CRA independent of any effect on the primordial pool. Oestrogen is, therefore, not a useful biomarker of damage to the primordial and early growing pool of follicles, with the clinical scenario being the frequent occurrence of amenorrhoea in women receiving chemotherapy who later recover and resume normal menstrual cycles – thus indicating that at least some of the ovarian reserve remains.

Many retrospective analyses have compared different chemotherapy regimens on the prevalence of amenorrhoea, as have large prospective analyses. A study by Petrek and colleagues is a very clear example of the acute impact of chemotherapy in women with breast cancer, with many developing CRA during and following treatment, with recovery of menstruation in some women following the end of treatment.[4] The overall prevalence of ongoing menses, however, subsequently declines as more women develop POI. This study also very strikingly shows the huge impact of the woman's age on the likelihood of ongoing menses after chemotherapy, varying from approximately 90% in women aged less than 35, to about 30% in women aged 40 and above. An important addition to these analyses is the demonstration that women who retain ovarian function after chemotherapy are at increased risk of an early menopause,[5] indicating the consequence of loss of much but not all of the ovarian reserve during chemotherapy.

3.3 BIOMARKERS OF THE OVARIAN RESERVE

There has long been interest in the development of accurate tools to detect the progressive loss of the ovarian reserve, which occurs with ageing, and indeed the incomplete loss from ovarian toxicity. This is in many ways analogous to the situation in assisted conception, where there is a need to distinguish between women with varying likely responses, from those with a poor or minimal response to superovulation, to those who will show an excessive response. There is a very substantial literature on assessment of the ovarian reserve in this context,[6–8] with initial studies exploring the value of FSH measurement, either basally or after a stimulatory test. FSH is clearly of value in diagnosing the menopause or POI, but it is of lesser value in detecting women of a low ovarian reserve. This is to be expected given its physiological role in selection of the dominant follicle and the later stages of antral follicle growth, thus being central to the ovulatory function of the ovary rather than the regulation of the early stages of follicle growth.

Current interest focuses around the use of measurement of circulating anti-Müllerian hormone (AMH), and ultrasound assessment of the number of small (generally 2–10 mm diameter) antral follicles: the antral follicle count (AFC). In the context of assisted reproduction, both are of comparable value in identifying both poor responders and over-responders.[8–10] The term "ovarian reserve" is generally used in assisted reproduction to indicate the response to ovarian stimulation, although it is important to recognize that the follicles that will be recruited by exogenous FSH are already at a relatively advanced stage of growth, i.e., at least at early antral stages. This analysis, therefore, does not directly reflect the true ovarian reserve, i.e., the primordial pool, although in adult women in normal healthy circumstances there is a relationship between the two.[11] Different relationships hold, however, in childhood and adolescence, where AMH rises despite a falling ovarian reserve,[12] although individual girls retain relative AMH levels across that time period.[13] AMH can also be influenced by external factors, most importantly in this context health status, with reductions in AMH in proportion to markers of general ill-health, including pyrexia and anaemia,[14] and during administration of hormonal contraception.[15]

AMH is produced by the granulosa cells of growing follicles, with expression initiated as soon as the follicle starts to grow, but it is not produced by the non-growing pool.[16] Expression continues through to the antral stages, but shows a rapid decline at a follicle diameter of about 8−10 mm diameter, i.e., at the time the follicle is potentially selected for dominance.[17] There is very little AMH production by larger antral follicles beyond this stage, which is of very substantial clinical benefit as it means that there is very little variation within the menstrual cycle. Thus, blood sampling can be taken at any time, in contrast to the measurement of FSH, which needs to be taken in the early follicular phase. Most of the AMH in the circulation is thought to come from the pool of small antral follicles detected by ultrasound, and there is generally a very close relationship between AMH and AFC measurements. Their value in assisted reproduction is strengthened by the fact that it is these same follicles that will be stimulated by exogenous FSH. At present, there remains no direct way of assessing the primordial follicle pool without using destructive histological analysis, although this remains an ideal. The limited activity of primordial follicles is a major hurdle in the development of such a biomarker. AMH and AFC have both been used to assess ovarian toxicity of chemotherapy regimens and it is clear that they are of value in this regard.[18−20] It is, however, very important to recognize that what they are and are not measuring.

3.4 AMH AND DETERMINATION OF GONADOTOXICITY

3.4.1 After Chemotherapy

AMH was first shown to be of potential value in assessing ovarian toxicity in cancer patients in a study of young adult women with regular menstrual cycles who were survivors of childhood cancers.[18] AMH but not FSH or inhibin B was significantly reduced in cancer survivors compared to age-matched controls. Ultrasound was also used in that study, and showed a reduced ovarian volume in cancer survivors having spontaneous cycles, although AFC was not reduced. Several studies have subsequently confirmed reduced AMH in survivors of adult and childhood cancer, with some of these finding clear relationships between the degree of loss of AMH and the gonadotoxicity of the treatment involved. Thus, women who have received alkylating agent-based therapy have lower AMH levels than those who received less toxic therapy[21–24]: this has also been confirmed in prospective studies.[25,26] A dose response is also apparent, with progressively lower AMH concentrations in women treated with more cycles of alkylating-based therapy for lymphoma.[21] Similarly, abdomino-pelvic irradiation and stem cell transplantation have very marked effects on AMH levels.[23] AMH has also been used to show the marked gonadotoxicity of radiotherapy involving the ovaries. In general, comparable data have been obtained using AFC,[15,20,27,28] although the literature on this is substantially sparser.

3.4.2 Prospective Analyses

There are now a growing number of studies that have prospectively assessed AMH as a marker of gonadotoxicity, thus making it possible to use pretreatment samples to show the fall in AMH in individuals, rather than requiring age-matched controls. This has confirmed the rapid loss of AMH during chemotherapy, reflecting the acute loss of growing follicles.[19] The fall in AMH appears more rapid and complete than changes in oestradiol and inhibin B, at least with the chemotherapy regimens (for early breast cancer) in that study. Further studies are required to dissect how useful this might be to reflect toxicity of different regimes on different stages of follicle growth. Prospective studies can also investigate the potential recovery of ovarian function after chemotherapy, and this has been clearly demonstrated to reveal differences between different therapeutic regimens for lymphoma.[25] Women with higher pretreatment AMH also were found to have more rapid recovery than those with lower pretreatment AMH.[26]

It is now becoming clear that pretreatment AMH concentrations can be used to predict the likelihood of long-term ovarian function after chemotherapy. This is in keeping with the concept that women with a higher pretreatment ovarian reserve are more likely to have more follicles that will survive treatment and that they will, therefore, have sufficient ovarian reserve to support clinical ovarian activity, i.e., continuation of or resumption of menses (Figure 3.1). This has been

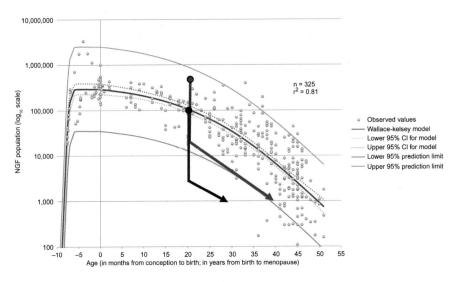

Figure 3.1 **Illustration of the impact of differences in the ovarian reserve on long-term ovarian function in two hypothetical cases.** *The background graph shows the decline in the ovarian reserve (number of non-growing follicles) against age, reproduced from*[2] *(doi:10.1371/journal.pone.0008772). The red and blue circles denote high and normal ovarian reserve respectively in two women aged 20 at the time of cancer diagnosis. The vertical line from each then shows the effect of loss of approximately 95% of the ovarian reserve, with the sloping arrowed line (in parallel to the normal rate of change) showing the predicted impact on age of menopause. In this example, the woman whose ovarian reserve is close to the mean will have her menopause 10 years after diagnosis, whereas the woman with a high ovarian reserve will have her menopause 20 years after diagnosis.*

most clearly shown thus far in women with breast cancer,[29–31] and data are required to confirm its relevance in other conditions, with their different associated treatment regimens and age profile of patients. While AMH appears of particular value in this context, it is important to recognize that age also has a role in this pretreatment assessment, as it remains an independent predictor.[31,32] This is analogous to data emerging on the interaction between age and AMH in predicting the menopause in normal women.[33] Ovarian function in these studies is of value in showing the endocrine activity of the ovary, but there are no equivalent data in relation to fertility. This is of significant importance as surveys of women treated for cancer, either as adults or in childhood, show a high prevalence of subfertility in those with ongoing ovarian activity[34,35]; the cause of this is unclear.

3.4.3 Measurement of AMH

The sensitivity of available assays for AMH has been an issue particularly in the analysis of post-chemotherapy ovarian function: essentially AMH has been undetectable in many women, despite ongoing ovarian activity.[19] This is analogous to the finding that AMH becomes undetectable approximately 5 years before the menopause in normal women.[36] However, new assays with markedly improved sensitivity are now becoming available, allowing the detection and assessment of ovarian activity in more women post-chemotherapy than previously possible.[37,38] Automated assays have recently become available from both Beckman Coulter and Roche Diagnostics, both of which offer high sensitivity as well as improved reproducibility compared to the widely used earlier manual assays. These are likely to be of considerable value in clinically assessing ovarian toxicity.

3.4.4 AMH in Paediatric Oncology

The assessment of gonadotoxicity in prepubertal girls is a particular challenge. AMH may well also be of value in this context, as it is readily detectable in girls of all ages, with its concentration showing a general increase through childhood and adolescence.[12] Indeed, peak serum concentrations occur at approximately age 24. In a prospective study AMH was found to fall during successive chemotherapy cycles in a population of girls and adolescents with various diagnoses, becoming undetectable in many at the end of therapy.[39] Most importantly, however, there was a clear distinction in the pattern of recovery between those receiving the most gonadotoxic therapy (high-dose alkylating agents or abdominopelvic radiotherapy) who showed no recovery in AMH, with it remaining undetectable during prolonged follow-up (Figure 3.2). In contrast, girls receiving less toxic therapy showed recovery to, on average, concentrations similar to those pre-treatment or even a little higher, perhaps reflecting the progressive rise in AMH through childhood. It may well be that post-treatment AMH will be of value in individual assessment of the potential for compromise of future reproductive lifespan, but this remains to be demonstrated. It does seem very likely, however, that those with continuing undetectable AMH are likely to require treatment to induce puberty at the appropriate age.

3.4.5 AMH and Prediction of Reproductive Lifespan and Fertility

There is increasing evidence that AMH may be of value in predicting age at natural menopause.[33,40,41] The analyses published thus far show wide confidence intervals and it is unclear at present how precise and, therefore, clinically useful to the individual woman a single

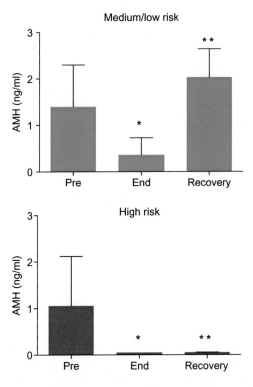

Figure 3.2 **Anti-Müllerian hormone (AMH) concentrations before treatment (Pre), at the end of treatment, and at more than 6 months' recovery after treatment in girls and adolescents.** *Patients were stratified according to predicted risk of gonadotoxicity. Median ± interquartile range, n = 9, high risk; n = 13, medium/low risk. *P < 0.01 vs. pretreatment in both groups, also between risk groups at end of treatment; **P < 0.01 vs. end of treatment in low/ medium risk group only, also between risk groups during recovery period.* Reprinted with permission from[39].

measurement may be. No doubt future studies will refine this, and explore multiple measurements to individualize rates of decline. There is a clear parallel between these studies and those exploring ovarian function after gonadotoxic therapy, and studies exploring the predictive value of post-chemotherapy AMH in subsequent reproductive lifespan are required, including assessment of fertility. In this context, there are few data exploring AMH and natural fecundity, with one study in younger women showing that those with an AMH in the lowest quintile had similar time to pregnancy to those with normal AMH levels,[42] whereas a study in women mostly in their 30s showed that those who had lower AMH levels had a lower likelihood of conceiving during the period of follow-up.[43] The available data in young cancer survivors indicates that even women with low AMH concentrations have a good chance of natural conception.[44] Again, age is likely to be a critically important cofactor here.

3.5 CONCLUSION

There is now a growing literature on the use of markers of the ovarian reserve to assess both the acute gonadotoxicity of different chemotherapy regimens and relationships between a woman's pre-existing ovarian function and her post-chemotherapy status. Substantial further work with long-term studies, including fertility as an outcome, is required to develop the maximum clinical utility of this approach, but it is clear that it has value both clinically and scientifically in assessing ovarian gonadotoxicity.

ACKNOWLEDGEMENT

The author's studies in this area have been supported by MRC grants WBS U.1276.00.002.00001 and G1100357.

REFERENCES

1. McGee EA, Hsueh AJ. Initial and cyclic recruitment of ovarian follicles. *Endocr Rev.* 2000;21(2):200−214.

2. Wallace WH, Kelsey TW. Human ovarian reserve from conception to the menopause. *PLoS One.* 2010;5(1):e8772.

3. Torino F, Barnabei A, De Vecchis L, et al. Chemotherapy-induced ovarian toxicity in patients affected by endocrine-responsive early breast cancer. *Crit Rev Oncol Hematol.* 2014;89(1):27−42.

4. Petrek JA, Naughton MJ, Case LD, et al. Incidence, time course, and determinants of menstrual bleeding after breast cancer treatment: a prospective study. *J Clin Oncol.* 2006;24 (7):1045−1051.

5. Partridge A, Gelber S, Gelber RD, Castiglione-Gertsch M, Goldhirsch A, Winer E. Age of menopause among women who remain premenopausal following treatment for early breast cancer: long-term results from International Breast Cancer Study Group Trials V and VI. *Eur J Cancer.* 2007;43(11):1646−1653.

6. Broekmans FJ, Kwee J, Hendriks DJ, Mol BW, Lambalk CB. A systematic review of tests predicting ovarian reserve and IVF outcome. *Hum Reprod Update.* 2006;12(6):685−718.

7. Broer SL, van Disseldorp J, Broeze KA, et al. Added value of ovarian reserve testing on patient characteristics in the prediction of ovarian response and ongoing pregnancy: an individual patient data approach. *Hum Reprod Update.* 2013;19(1):26−36.

8. Broer SL, Dolleman M, Opmeer BC, Fauser BC, Mol BW, Broekmans FJ. AMH and AFC as predictors of excessive response in controlled ovarian hyperstimulation: a meta-analysis. *Hum Reprod Update.* 2011;17(1):46−54.

9. Broer SL, Dolleman M, van Disseldorp J, et al. Group, I-ES Prediction of an excessive response in *in vitro* fertilization from patient characteristics and ovarian reserve tests and comparison in subgroups: an individual patient data meta-analysis. *Fertil Steril.* 2013;100 (2):420−429:e7.

10. La Marca A, Sunkara SK. Individualization of controlled ovarian stimulation in IVF using ovarian reserve markers: from theory to practice. *Hum Reprod Update.* 2014;20(1):124−140.

11. Hansen KR, Hodnett GM, Knowlton N, Craig LB. Correlation of ovarian reserve tests with histologically determined primordial follicle number. *Fertil Steril.* 2011;95:170−175.

12. Kelsey TW, Wright P, Nelson SM, Anderson RA, Wallace WH. A validated model of serum anti-Müllerian hormone from conception to menopause. *PLoS One.* 2011;6(7):e22024.

13. Hagen CP, Aksglaede L, Sorensen K, et al. Individual serum levels of anti-Müllerian hormone in healthy girls persist through childhood and adolescence: a longitudinal cohort study. *Hum Reprod.* 2012;27(3):861−866.

14. van Dorp W, van den Heuvel-Eibrink MM, de Vries AC, et al. Decreased serum anti-Müllerian hormone levels in girls with newly diagnosed cancer. *Hum Reprod.* 2014;29 (2):337−342.

15. Johnson LN, Sammel MD, Dillon KE, Lechtenberg L, Schanne A, Gracia CR. Antimüllerian hormone and antral follicle count are lower in female cancer survivors and healthy women taking hormonal contraception. *Fertil Steril.* 2014;102(3):774−781:e3.

16. Weenen C, Laven JS, Von Bergh AR, et al. Anti-Müllerian hormone expression pattern in the human ovary: potential implications for initial and cyclic follicle recruitment. *Mol Hum Reprod.* 2004;10(2):77−83.

17. Jeppesen JV, Anderson RA, Kelsey TW, et al. Which follicles make the most anti-Müllerian hormone in humans? Evidence for an abrupt decline in AMH production at the time of follicle selection. *Mol Hum Reprod.* 2013;19:519−527.

18. Bath LE, Wallace WH, Shaw MP, Fitzpatrick C, Anderson RA. Depletion of ovarian reserve in young women after treatment for cancer in childhood: detection by anti-Müllerian hormone, inhibin B and ovarian ultrasound. *Hum Reprod.* 2003;18(11):2368−2374.

19. Anderson RA, Themmen APN, Al Qahtani A, Groome NP, Cameron DA. The effects of chemotherapy and long-term gonadotrophin suppression on the ovarian reserve in premenopausal women with breast cancer. *Human Reprod.* 2006;21(10):2583−2592.

20. Lutchman Singh K, Muttukrishna S, Stein RC, et al. Predictors of ovarian reserve in young women with breast cancer. *Br J Cancer.* 2007;96(12):1808−1816.

21. van Beek RD, van den Heuvel-Eibrink MM, Laven JS, et al. Anti-Müllerian hormone is a sensitive serum marker for gonadal function in women treated for Hodgkin's lymphoma during childhood. *J Clin Endocrinol Metab.* 2007;92(10):3869−3874.

22. Rosendahl M, Andersen CY, Ernst E, et al. Ovarian function after removal of an entire ovary for cryopreservation of pieces of cortex prior to gonadotoxic treatment: a follow-up study. *Hum Reprod.* 2008;23(11):2475−2483.

23. Lie Fong S, Lugtenburg PJ, Schipper I, et al. Anti-müllerian hormone as a marker of ovarian function in women after chemotherapy and radiotherapy for haematological malignancies. *Hum Reprod.* 2008;23(3):674−678.

24. Lie Fong S, Laven JS, Hakvoort-Cammel FG, et al. Assessment of ovarian reserve in adult childhood cancer survivors using anti-Müllerian hormone. *Hum Reprod.* 2009;24(4):982−990.

25. Decanter C, Morschhauser F, Pigny P, Lefebvre C, Gallo C, Dewailly D. Anti-Müllerian hormone follow-up in young women treated by chemotherapy for lymphoma: preliminary results. *Reprod Biomed Online.* 2010;20(2):280−285.

26. Dillon KE, Sammel MD, Prewitt M, et al. Pretreatment antimüllerian hormone levels determine rate of posttherapy ovarian reserve recovery: acute changes in ovarian reserve during and after chemotherapy. *Fertil Steril.* 2013;99:477−483.

27. Su HI, Chung K, Sammel MD, Gracia CR, DeMichele A. Antral follicle count provides additive information to hormone measures for determining ovarian function in breast cancer survivors. *Fertil Steril.* 2011;95(5):1857−1859.

28. Partridge AH, Ruddy KJ, Gelber S, et al. Ovarian reserve in women who remain premenopausal after chemotherapy for early stage breast cancer. *Fertil Steril.* 2010;94(2):638−644.

29. Anders C, Marcom PK, Peterson B, et al. A pilot study of predictive markers of chemotherapy-related amenorrhea among premenopausal women with early stage breast cancer. *Cancer Invest.* 2008;26(3):286−295.

30. Anderson RA, Cameron DA. Pretreatment serum anti-müllerian hormone predicts long-term ovarian function and bone mass after chemotherapy for early breast cancer. *J Clin Endocrinol Metab.* 2011;96(5):1336−1343.

31. Anderson RA, Rosendahl M, Kelsey TW, Cameron DA. Pretreatment anti-Müllerian hormone predicts for loss of ovarian function after chemotherapy for early breast cancer. *Eur J Cancer.* 2013;49(16):3404−3411.

32. Ruddy KJ, O'Neill A, Miller KD, et al. Biomarker prediction of chemotherapy-related amenorrhea in premenopausal women with breast cancer participating in E5103. *Breast Cancer Res Treat.* 2014;144(3):591−597.

33. Freeman EW, Sammel MD, Lin H, Gracia CR. Anti-müllerian hormone as a predictor of time to menopause in late reproductive age women. *J Clin Endocrinol Metab.* 2012;97 (5):1673−1680.

34. Letourneau JM, Ebbel EE, Katz PP, et al. Acute ovarian failure underestimates age-specific reproductive impairment for young women undergoing chemotherapy for cancer. *Cancer.* 2012;118:1933−1939.

35. Barton SE, Najita JS, Ginsburgh ES, et al. Infertility, infertility treatment, and achievement of pregnancy in female survivors of childhood cancer: a report from the Childhood Cancer Survivor Study cohort. *Lancet Oncol.* 2013;14(9):873−881.

36. Sowers MR, Eyvazzadeh AD, McConnell D, et al. Anti-Müllerian hormone and inhibin B in the definition of ovarian aging and the menopause transition. *J Clin Endocrinol Metab.* 2008;93(9):3478−3483.

37. Chai J, Howie AF, Cameron DA, Anderson RA. A highly-sensitive anti-Müllerian hormone assay improves analysis of ovarian function following chemotherapy for early breast cancer. *Eur J Cancer.* 2014;50(14):2367−2374.

38. Decanter C, Peigne M, Mailliez A, et al. Toward a better follow-up of ovarian recovery in young women after chemotherapy with a hypersensitive antimüllerian hormone assay. *Fertil Steril.* 2014;102(2):483−487.

39. Brougham MF, Crofton PM, Johnson EJ, Evans N, Anderson RA, Wallace WH. Anti-Müllerian hormone is a marker of gonadotoxicity in pre- and postpubertal girls treated for cancer: a prospective study. *J Clin Endocrinol Metab.* 2012;97:2059−2067.

40. Broer SL, Eijkemans MJ, Scheffer GJ, et al. Anti-Müllerian hormone predicts menopause: a long-term follow-up study in normoovulatory women. *J Clin Endocrinol Metab.* 2011;96 (8):2532−2539.

41. Tehrani FR, Solaymani-Dodaran M, Tohidi M, Gohari MR, Azizi F. Modeling age at menopause using serum concentration of anti-müllerian hormone. *J Clin Endocrinol Metab.* 2013;98(2):729−735.

42. Hagen CP, Vestergaard S, Juul A, et al. Low concentration of circulating anti-Müllerian hormone is not predictive of reduced fecundability in young healthy women: a prospective cohort study. *Fertil Steril.* 2012;98(6):1602−1608.

43. Steiner AZ, Herring AH, Kesner JS, et al. Antimüllerian hormone as a predictor of natural fecundability in women aged 30−42 years. *Obstet Gynecol.* 2011;117(4):798−804.

44. Hamre H, Kiserud CE, Ruud E, Thorsby PM, Fossa SD. Gonadal function and parenthood 20 years after treatment for childhood lymphoma: a cross-sectional study. *Pediatr Blood Cancer.* 2012;59(2):271−277.

CHAPTER 4

The Current Understanding of Clinical Data on Ovarian Toxicity from Cancer Treatment

Volkan Turan and Kutluk Oktay

Division of Reproductive Medicine and Laboratory of Molecular Reproduction & Fertility Preservation, Obstetrics and Gynecology, New York Medical College, Valhalla, NY, USA; Innovation Institute for Fertility Preservation and IVF, New York, NY, USA

4.1 INTRODUCTION

Chemotherapy has considerably improved survival of many cancer types; however, a patient's quality of life is influenced by long-term complications of chemotherapeutic regimens.[1,2] These complications include cardiac abnormalities, secondary malignancies, renal and hepatic impairment, and gonadal dysfunction.[3] Gonadal dysfunction may result in an early menopause and infertility in reproductive-age patients, which is associated with poorer quality of life.[4] Chemotherapy-induced gonadotoxicity is not limited to cancer patients but also includes patients with rheumatologic diseases, aplastic anaemia and systemic lupus erythematosus. Therefore, patients diagnosed with these conditions are also at risk of premature ovarian insufficiency (POI) after treatment.[5]

4.2 RISK FACTORS FOR PREMATURE OVARIAN INSUFFICIENCY AFTER CHEMOTHERAPY

Most long-term follow-up studies evaluating ovarian toxicity after adjuvant chemotherapy are retrospective and have used the menstrual pattern as the only surrogate marker.[6−9] The extent of gonadotoxic injury following chemotherapy is influenced by the age of the patient at the time of treatment, as well as the type, dose and duration of chemotherapy,[10,11] and the clinical degree of ovarian dysfunction, which varies from transient amenorrhoea to true menopause, may reflect the magnitude of gonadotoxicity.

Cancer Treatment and the Ovary. DOI: http://dx.doi.org/10.1016/B978-0-12-801591-9.00004-7

4.2.1 Age

Age-related differences are most likely due to a reduction of the ovarian reserve with ageing, which increases the risk of developing POI post-chemotherapy. Females less than 40 years of age exposed to chemotherapy agents have a 22–61% risk of developing amenorrhoea, and this rate increases to 66–95% in females older than 40 years.[12–14] Resumption of menstruation was higher in patients younger than 40 years compared to those older than 40 years.[6] In the National Surgical Adjuvant Breast and Bowel Project (NSABP) B-30 study, 708 patients diagnosed with breast cancer were treated with four cycles of doxorubicin and cyclophosphamide (AC) and four cycles of docetaxel. The resumption of menses occurred within 24 months in 45.3% of patients aged under age 40 years, 10.9% of those aged 40–50 and in 3.2% of patients aged over 50 years.[9]

4.2.2 Type of Chemotherapy

The type of chemotherapy is a strong determinant of POI (Figure 4.1). Alkylating agents carry the highest risk because these drugs are not cell cycle specific, meaning that both resting and developing follicles can be damaged.[15] Cyclophosphamide is the most commonly utilized

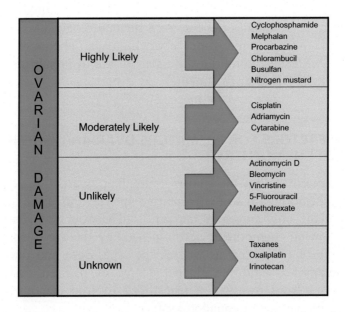

Figure 4.1 The risk of POI according to different chemotherapeutic agents.

agent in this group; it is frequently used to treat a wide range of cancers including lymphomas, leukaemia, neuroblastoma, retinoblastoma and breast carcinoma. Other clinical uses for cyclophosphamide include immunosuppressive therapy following organ transplants or as a treatment for connective tissue disorders. Cyclophosphamide leads to major disruptions in nucleic acid function and inhibits DNA and protein synthesis. Its use is associated with DNA crosslinking in granulosa cells, stromal fibrosis and capillary changes in the female reproductive system.[16] Gonadal toxicity with cyclophosphamide alone or in combination with other chemotherapeutics has been documented.[17–19] The authors reported previously that even a single dose of cyclophosphamide resulted in a drastic reduction in human primordial follicle reserve in a xenograft model.[17] Boumpas et al.[19] compared the effects of short (seven doses) and long (>15 doses) courses of cyclophosphamide treatment in 39 females aged under age 40 years diagnosed with systemic lupus erythematosus to evaluate the effect of number of doses of cyclophosphamide on menstruation and the incidence of sustained amenorrhoea was found to be higher in the long-course group (12.5 vs. 39%, p = 0.07).

Doxorubicin, a topoisomerase inhibitor antracyclin antibiotic, is widely integrated in a variety of cancer regimens, and acts by intercalating DNA. It is one of the most effective drugs for the treatment of solid tumours such as breast cancer. DNA damage and apoptotic cell death in human granulosa cells and primordial follicles, as well as a reduction in stromal blood flow, has been shown after in vivo administration of doxorubicin.[20,21] Depending on intensity, duration, and combination with other chemotherapy agents, doxorubicin treatment may result in POI in cancer patients.

Taxanes are relatively new chemotherapeutic agents that act on the cytoskeleton to disrupt microtubule function. They are commonly used to consolidate doxorubicin- and cyclophosphamide-based treatment in breast cancer patients.[22,23] Laboratory studies of taxane-induced ovarian toxicity are limited and conflicting. Tarumi et al.[24] suggested that the ovarian toxicity of paclitaxel is mild and does not involve the primordial follicles; however, in another study in rats, primordial follicle counts decreased following the administration of paclitaxel.[25] Recently, Lopes et al. demonstrated that docetaxel directly impairs the early stages of ovarian follicle development, but it has no direct effect on the primordial follicle reserve in mice.[26] Though the data are limited, clinically, taxanes do not appear to have a strong ovary-damaging effect.[27]

Cycle-specific agents — such as 5-fluorouracil, methotrexate and vinca alkaloids — appear to affect only the developing and mature follicles and hence cause transient amenorrhoea without damaging primordial follicles.[15]

4.2.3 Combined Treatment Protocols

The assessment of ovarian function after chemotherapeutic agent combinations in the treatment of different cancer types is, in most studies, based on the menstrual pattern. However, comparisons of chemotherapy-induced amenorrhoea rates and resumption of menstruation exhibit substantial variability due to differences in mean age, treatment dose, and follow-up duration.[28–31]

4.2.3.1 Breast Cancer

Breast cancer is the most frequent cancer diagnosed in females of reproductive age.[32] Because breast cancer in young females presents with a high prevalence of ductal infiltration, most of those patients are likely to undergo adjuvant systemic chemotherapy.[33] Although a cyclophosphamide, methotrexate and 5-fluorouracil regimen (CMF) is rarely used today in breast cancer, in a detailed analysis of the published studies with classic CMF, the incidence of amenorrhoea ranged from 21–71% in females younger than 40 years of age and 40–100% in older females.[2,13] The actual risk of POI associated with taxanes, which have been incorporated recently in the adjuvant setting for breast cancer, remains controversial. With AC-based regimens, the incidence of amenorrhoea ranged from 15–93% with or without taxanes.[34,35] A study by Tham et al.[25] included 191 premenopausal females with breast cancer and the rate of chemotherapy-induced amenorrhoea in patients who received AC followed by taxanes was significantly higher compared with AC alone (64 vs. 55%, p = 0.05). In addition, chemotherapy-induced amenorrhoea rates were higher in older than younger females (82 vs. 55%, p = 0.004). However, Davis et al.[24] found no difference in amenorrhoea rates between females receiving AC-based chemotherapy with or without consolidating docetaxel in review of 159 premenopausal females. The results of another study comparing a similar regimen, 5-fluoruracil, epirubicin and cyclophosphamide (FEC) with or without docetaxel, were consistent with the findings of Davis et al.[36] In this study, 154 premenopausal patients were included and 84 treated with six cycles of FEC and 70 with three cycles of FEC plus docataxel. The incidence of chemotherapy-induced amenorrhoea at the end of chemotherapy was similar in the two groups (93% vs. 92.8%).

Resumption of menstruation can occur early or be delayed 1–2 years from the initiation of chemotherapy-related amenorrhoea, depending on the type of treatment. Fornier et al.[34] investigated long-term amenorrhoea lasting >12 months in 166 females with breast carcinoma after adjuvant AC- and taxane-based chemotherapy. In this study 25 patients (15%) developed long-term amenorrhoea, and 141 patients (85%) resumed menstruation. In a large prospective study investigating the maintenance of menstrual bleeding in 466 patients with breast cancer that received AC, AC-T or CMF, females receiving CMF were less likely to menstruate after 6 months of follow-up compared to AC and AC-T regimens (23% with CMF vs. 68% with AC vs. 57% with AC-T, p = 0.002).[37] Berliere et al.[36] compared resumption of menses after FEC vs. FEC plus docetaxel; more patients with consolidating docataxel recovered menses in the year following the end of chemotherapy (35.5 vs. 23.7%, p = 0.01).

4.2.3.2 Hodgkin Lymphoma

Hodgkin lymphoma (HL) is a relatively rare cancer with an incidence of 2–3 per 100,000 females.[38] In HL, treatment protocols combining doxorubicin, bleomycin, vinblastine and dacarbazine (ABVD), without alkylating agents, rarely result in POI;[39,40] however, treatment protocols containing alkylating agents, especially procarbazine and cyclophosphamide, induce POI more often, varying from 20–85% depending on the protocol.[41] Whitehead et al.[42] investigated the effect of combination chemotherapy on ovarian function in females treated for HL. Ovarian function was studied in 44 adult females who previously received mechlorethamine, vincristine, procarbazine and prednisone (MOPP) for HL. The median age at treatment was 23 years, and the length of time between completion of treatment and study ranged from 6 months to 10 years (median, 30 months). Seventeen females (38%) developed amenorrhoea and 10 (22.7%) developed oligomenorrhoea. In another study by Schilsky et al.,[8] persistent amenorrhoea occurred in 11 of 24 patients (46%) treated with the MOPP protocol and the time from diagnosis to amenorrhoea was significantly shorter in older patients (p = 0.001). The German Hodgkin Lymphoma Study Group[7] analysed 405 females with HL younger than 40 years of age who were given dose-escalated BEACOPP (bleomycin, etoposide, doxorubicin, cyclophosphamide, vincristine, procarbazine and prednisolone), ABVD, or standard BEACOPP. After a median follow-up of 3.2 years, 51.4% of the females receiving eight cycles of dose-escalated BEACOPP had continuous amenorrhoea. Amenorrhoea was significantly more frequent after dose-escalated BEACOPP compared with ABVD or standard BEACOPP (p = 0.006). More females over the age of 30 treated for advanced-stage HL with four cycles of alternating COPP/ABVD or eight cycles of standard BEACOPP or eight cycles of dose-escalated BEACOPP experienced amenorrhoea compared with females younger than 30 years (66.7 vs. 40% with COPP/ABVD, 53.3 vs. 18.2% with standard BEACOPP, 95 vs. 51.4% with dose-escalated BEACOPP).

4.2.3.3 Non-Hodgkin Lymphoma

The incidence of Non-Hodgkin lymphoma (NHL) in females has been reported as 7–9 per 100,000.[38] Most treatment regimens for NHL include alkylating agents. Ellis *et al.*[43] evaluated a cohort of 36 females younger than 40 years of age with aggressive NHL treated with mostly cyclophosphamide, adriamycin, oncovine and prednisone (CHOP). During treatment, 18 patients (50%) had amenorrhoea, six (17%) had irregular menstrual cycles and 12 (33%) continued their regular cycles. All but two females resumed menses in the first complete remission. Dann *et al.*[44] analysed 13 females with NHL less than 40 years of age, treated with four cycles of intensified CHOP (cyclophosphamide 2000–3000 mg/m^2 per cycle, doxorubicin 50 mg/m^2, vincristine 1.4 mg/m^2, prednisone 100 mg/day) with a median follow-up of 70 months and reported that 12 of 13 patients (92%) experienced recovery of ovarian function.

4.2.3.4 Leukaemia

Leukaemia is the most commonly diagnosed cancer in children, and acute lymphoblastic leukaemia (ALL) accounts for 80% of all leukaemias.[45] Modern chemotherapy protocols of childhood ALL or acute myeloid leukaemia are unlikely to affect ovarian function in many patients due to the absence or lower dose of alkylating agents when haematopoietic stem cell transplantation (HSCT) is not required.[46]

However, HSCT requires preconditioning regimens such as alkylating agents with or without total body irradiation, resulting in infertility and POI.[47]

4.2.3.5 Role of Gonadotropin-Releasing Hormone Analogues for the Preservation of Ovarian Function

In several studies, it was shown that when gonadotropin-releasing hormone (GnRH) analogues were given as adjuvant therapy to protect the ovaries from the deleterious effect of chemotherapy, the reported rates of amenorrhoea were lower compared with chemotherapy alone.[48,49] However, no study has demonstrated a beneficial effect regarding fertility recovery.[50,51] Since the current data indicate that infertility is increased after chemotherapy, even if menstrual cycles are resumed,[52] further research is required to investigate the role of GnRH agonist administration in fertility preservation. This topic is covered in detail in Chapter 8.

4.3 HOW DO CHEMOTHERAPEUTICS DAMAGE THE OVARY?

Chemotherapy-induced ovarian damage is associated with a decrease in the follicular pool, impaired blood vasculature and stromal fibrosis.[10,20,53] The existing molecular and *in vivo* evidence show that chemotherapy treatments reduce the ovarian reserve by directly damaging DNA in human primordial follicle oocytes.[20] Based on molecular studies, both cyclophosphamide and doxorubicin affect primordial follicle reserve, which shortens the reproductive lifespan.[17,20] Both agents are toxic against both the primordial follicle population and those that are in growth phase. Both cyclophosphamide and doxorubicin induce double-strand DNA breaks (DSBs) in oocytes and granulosa cells, which likely triggers a germ cell-specific p53 family member transcription factor TAP-63-mediated apoptotic death. However, there is some indirect evidence that some oocytes may avoid apoptotic cell death when the ataxia telangiectasia mutated (ATM)-mediated DNA DSB repair pathway is activated, thus repairing damaged DNA[54] (Figure 4.2).

Others have suggested alternative mechanisms based on work in rodents.[55] One idea put forward was that chemotherapy induced massive activation of primordial follicles ("burn out") via the PI3K/PTEN/Akt pathway, in mice. However, this theory does not explain the

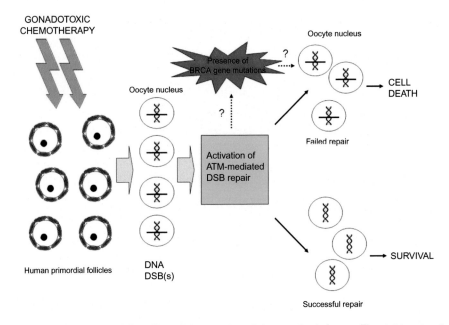

*Figure 4.2 **The mechanism of chemotherapy-induced ovarian follicle reserve loss in humans.** The administration of chemotherapy agents such as cyclophosphamide and doxorubicin results in DNA double-strand breaks (DSBs) in primordial follicle oocyte and activation of ataxia telangiectasia mutated (ATM)-mediated DNA DSB repair pathway, which includes the BRCA genes. Oocytes with sufficient DNA repair ability may survive this genotoxic stress. Hypothetically, variations in DNA repair efficiency, epigenetic changes or single-gene mutations such as the one in BRCA genes may hamper successful repair, rendering susceptibility to chemotherapy-induced damage.*

increase in malformations of offspring when mouse primordial follicles were exposed to cyclophosphamide,[56] which we have shown to cause DNA DSBs in human primordial follicle oocytes.[54] In addition, it was shown that chemotherapy-induced DNA breaks in primordial follicles can even have transgenerational effects.[57] If the main cause of follicle death was due to the activation of primordial follicles and hence the later impact of chemotherapeutic agents on the more sensitive, now growing follicles, one would not expect to see transgenerational effects or the late malformations that have been reported.[57,58] While a transient increase in follicle growth initiation may contribute to follicle loss, as has been reported in rodents when ovarian reserve was acutely reduced by unilateral oophorectomy,[59] it is not known if this reaches proportions to explain the significant ovarian damage caused by chemotherapeutic agents in humans.

Granulosa cells are somatic cells that surround the oocyte and proliferate during follicle maturation, and may be another potential target of chemotherapeutic agents. Raz et al.[60] incubated human ovarian cortical slices from premenopausal females with medium containing cyclophosphamide for 2–48 h, and found damage in granulosa cell nuclei and follicular basement membranes. Morgan et al.[61] reported that doxorubicin preferentially damaged the granulosa cells, whereas cisplatin showed more oocyte-specific damage in mouse ovaries. The authors showed previously in a model of human fetal ovary pieces xenografted into severe combined immunodeficient mice that all or nearly all apoptotic follicles had an apoptotic oocyte with or without the pregranulosa cell layer that were also positive after cyclophosphamide treatment.[17] Since granulosa cells can be replenished, apoptosis of the oocyte may be considered more lethal to ovarian reserve.

Ovarian stromal tissue has also been shown to be vulnerable to chemotherapy in human studies.[20,53,62] Meirow et al.[53] investigated stromal damage in human ovaries after exposure to chemotherapy in cancer patients. In this study of 35 cancer patients (mean age; 28.7 years), 17 were previously exposed to chemotherapy and 18 were not. After histopathological evaluation, focal cortical fibrosis and blood vessel damage were detected in ovaries of patients previously exposed to chemotherapy. Marcello et al.[62] also reported severe signs of stromal fibrosis and capillary changes in ovarian biopsy specimens of 10 girls who underwent treatment for ALL. The authors previously assessed ovarian samples from 26 patients who were undergoing ovarian tissue cryopreservation for fertility preservation to detect stromal function via measuring in vitro oestradiol levels, and found that regardless of whether they include an alkylating agent, most chemotherapy regimens can have detrimental effects on ovarian stromal function.[16] The authors have also shown in both in vivo xenograft and in vitro organ culture models that gonadotoxic treatments cause microvascular damage, resulting in tissue hypoxia.[20] This may further contribute to the complex damage imparted by cancer treatments on ovarian function.

It has been shown that the ceramide-induced death pathway plays a role in the apoptotic death of mouse oocytes in response to radiation- and chemotherapy-induced damage. Sphingosine-1-phosphate (S1P) is a naturally occurring byproduct of that pathway that can negate the apoptotic consequences of the activation.[63–65] Numerous studies in rodent models have confirmed that the exogenous administration of S1P protects oocytes against chemotherapy- and radiotherapy-induced death.[63,64] The authors recently showed that S1P can block the human apoptotic follicle death induced by doxorubicin and cyclophosphamide in a human ovarian xenograft model.[65] Moreover, the research indicates that S1P does not interfere with the effectiveness of chemotherapy while protecting human ovarian reserve against chemotherapy.[66]

4.4 THE RELEVANCE OF *BRCA* MUTATIONS TO OVARIAN DAMAGE AFTER CHEMOTHERAPY

The authors have previously shown that women with breast cancer who are *BRCA1* carriers may display low response to ovarian stimulation.[67] *BRCA1* and *BRCA2* genes are key members of the ATM-mediated DNA DSB repair pathway and their detrimental mutations can put females at higher risk of developing breast, ovarian and other cancers.[68] The authors found that the expressions of the key DNA DSB repair genes, including especially the *BRCA1* gene, decline with age, rendering primordial follicle oocytes more susceptible to accumulating DNA DSBs with age. This then may trigger accelerated clearance of damaged primordial follicles via an apoptotic mechanism.[54] *BRCA* mutation carriers have been reported to experience menopause earlier compared to females without such a mutation.[69−72] While two studies showed that both *BRCA1* and *BRCA2* carriers experience early menopause compared to controls,[71,72] Rzepka-Gorska *et al.* found that *BRCA1* mutations are associated with earlier menopause.[70] Given that gonadotoxic chemotherapy drugs are severe DNA-damaging agents and trigger a repair response in oocytes via the ATM-mediated DSB repair pathway, which includes *BRCA* genes, females with *BRCA* mutations may be more susceptible to chemotherapy-induced ovarian follicle loss and enter menopause earlier compared to patients without mutations[54,71] (see Figure 4.2). However, this is currently speculative and future prospective studies are needed to study this hypothesis.

The authors recently showed that females with *BRCA1* mutations have lower ovarian reserve based on anti-Müllerian hormone (AMH) levels compared to controls without mutations and patients with *BRCA2* mutations.[67] In a recent study, Wang *et al.* reported that *BRCA1* mutation carriers have a significant decrease in AMH levels compared with controls after adjusting for age and body mass index (0.53 vs. 1.05 ng/ml; $p = 0.026$); however, no significant differences were found between *BRCA2* carriers and controls or between *BRCA1* and *BRCA2* carriers.[73] Earlier age-related decline in function of the *BRCA1* gene compared to *BRCA2* gene seems to explain the differences observed in ovarian reserve between women with *BRCA1* and *BRCA2* mutations. The significant decline in the function of the normal *BRCA2* allele happens very late or past reproductive ages, hence it does not significantly affect reproductive function. However, because the intact allele is subjected to environmental and epigenetic changes, *BRCA* gene function may decline at much earlier ages in certain populations or individuals, explaining the variation in the age of onset for breast and ovarian cancers in *BRCA* mutation carriers. In addition, the most severely affected individuals are likely being removed from the population of reproductive women due to early development of ovarian cancer and/or risk reducing salpingo-oophorectomy; therefore, small retrospective studies may not be able to detect differences between *BRCA* mutation carriers and non-carriers, as well those affected versus non-affected. Therefore, while it is possible that women with *BRCA1* mutations may be more vulnerable to chemotherapy-induced POI,[74] prospectively designed studies using more sensitive ovarian markers are required to evaluate this hypothesis.

4.5 CONCLUSION

Chemotherapeutic agents often have a negative impact on the gonads of reproductive-aged females and the extent of injury is influenced by the age of the patient and the type, dose and duration of chemotherapy. There is clear evidence that gonadotoxic chemotherapy agents can significantly impair DNA integrity by causing DNA DSBs in primordial follicle oocytes. *BRCA* and other ATM-mediated DNA repair genes may have a special role in the repair of chemotherapy-induced DNA DSBs. Future laboratory and prospective large clinical trials are needed to further elucidate the mechanisms of damage and repair in human ovarian oocytes in response to chemotherapy.

REFERENCES

1. Siegel R, Naishadham D, Jemal A. Cancer statistics, 2013. *CA Cancer J Clin.* 2013;63(1):11–30.

2. Fleischer RT, Vollenhoven BJ, Weston GC. The effects of chemotherapy and radiotherapy on fertility in premenopausal women. *Obstet Gynecol Surv.* 2011;66(4):248–254.

3. Leung W, Hudson MM, Strickland DK, et al. Late effects of treatment in survivors of childhood acute myeloid leukaemia. *J Clin Oncol.* 2000;18(18):3273–3279.

4. Turan V, Oktay K. Sexual and fertility adverse effects associated with chemotherapy treatment in women. *Expert Opin Drug Saf.* 2014;13(6):775−783.

5. Sonmezer M, Oktay K. Fertility preservation in female patients. *Hum Reprod Update.* 2004;10(3):251−266.

6. Bines J, Oleske DM, Cobleigh MA. Ovarian function in premenopausal women treated with adjuvant chemotherapy for breast cancer. *J Clin Oncol.* 1996;14(5):1718−1729.

7. Behringer K, Breuer K, Reineke T, et al. Secondary amenorrhea after Hodgkin's Lymphoma is influenced by age at treatment, stage of disease, chemotherapy regimen, and the use of oral contraceptives during therapy: a report from the German Hodgkin's Lymphoma Study Group. *J Clin Oncol.* 2005;23(30):7555−7564.

8. Schilsky RL, Sherins RJ, Hubbard SM, Wesley MN, Young RC, DeVita VT. Long-term follow-up of ovarian function in women treated with MOPP chemotherapy for Hodgkin's disease. *Am J Med.* 1981;71(4):552−556.

9. Swain SM, Land SR, Ritter MW, et al. Amenorrhea in premenopausal women on the doxorubicin-and-cyclophosphamide followed by docetaxel arm of NSABP B-30 trial. *Breast Cancer Res Treat.* 2009;113(2):315−320.

10. Morgan S, Anderson RA, Gourley C, Wallace WH, Spears N. How do chemotherapeutic agents damage the ovary? *Hum Reprod Update.* 2012;18(5):525−535.

11. Rose DP, Davis TE. Ovarian function in patients receiving adjuvant chemotherapy for breast cancer. *Lancet.* 1977;1(8023):1174−1176.

12. Torino F, Barnabei A, De Vecchis L, et al. Chemotherapy-induced ovarian toxicity in patients affected by endocrine-responsive early breast cancer. *Crit Rev Oncol Hematol.* 2014;89(1):27−42.

13. Oktay K, Sönmezer M. Chemotherapy and amenorrhea: risks and treatment options. *Curr Opin Obstet Gynecol.* 2008;20(4):408−415.

14. Minton SE, Munster PN. Chemotherapy-induced amenorrhea and fertility in women undergoing adjuvant treatment for breast cancer. *Cancer Control.* 2002;9(6):466−472.

15. Rodriguez-Wallberg KA, Oktay K. Fertility preservation and pregnancy in women with and without *BRCA* mutation-positive breast cancer. *Oncologist.* 2012;17(11):1409−1417.

16. Oktem O, Oktay K. Quantitative assessment of the impact of chemotherapy on ovarian follicle reserve and stromal function. *Cancer.* 2007;110(10):2222−2229.

17. Oktem O, Oktay K. A novel ovarian xenografting model to characterize the impact of chemotherapy agents on human primordial follicle reserve. *Cancer Res.* 2007;67(21): 10159−10162.

18. Byrne J, Fears TR, Gail MH, et al. Early menopause in long term survivors of cancer during adolescence. *Am J Obstet Gynecol.* 1992;166(3):788−793.

19. Boumpas DT, Austin III HA, Vaughan EM, Yarboro CH, Klippel JH, Balow JE. Risk for sustained amenorrhea in patients with systemic lupus erythematosus receiving intermittent pulse cyclophosphamide therapy. *Ann Intern Med.* 1993;119(5):366−369.

20. Soleimani R, Heytens E, Darzynkiewicz Z, Oktay K. Mechanisms of chemotherapy-induced human ovarian aging: double strand DNA breaks and microvascular compromise. *Aging.* 2011;3(8):782−793.

21. Ben-Aharon I, Bar-Joseph H, Tzarfaty G, et al. Doxorubicin induced ovarian toxicity. *Reprod Biol Endocrinol.* 2010;4:8−20.

22. Davis AL, Litus M, Mintzer DM. Chemotherapy-induced amenorrhea from adjuvant breast cancer treatment: the effect of the addition of taxanes. *Clin Breast Cancer.* 2005;6(5): 421−424.

23. Tham YL, Sexton K, Weiss H, Elledge R, Friedman LC, Kramer R. The rates of chemotherapy-induced amenorrhea in patients treated with adjuvant doxorubicin and cyclophosphamide followed by a taxane. *Am J Clin Oncol.* 2007;30(2):126–132.

24. Tarumi W, Suzuki N, Takahashi N, et al. Ovarian toxicity of paclitaxel and effect on fertility in the rat. *J Obstet Gynaecol Res.* 2009;35(3):414–420.

25. Yucebilgin MS, Terek MC, Ozsaran A, et al. Effect of chemotherapy on primordial follicular reserve of rat: An animal model of premature ovarian failure and infertility. *Aust N Z J Obstet Gynaecol.* 2004;44(1):6–9.

26. Lopes F, Smith R, Anderson RA, Spears N. Docetaxel induces moderate ovarian toxicity in mice, primarily affecting granulosa cells of early growing follicles. *Mol Hum Reprod.* 2014;20(10):948–959.

27. Reh A, Oktem O, Oktay K. Impact of breast cancer chemotherapy on ovarian reserve: a prospective observational analysis by menstrual history and ovarian reserve markers. *Fertil Steril.* 2008;90(5):1635–1639.

28. Han HS, Ro J, Lee KS, et al. Analysis of chemotherapy-induced amenorrhea rates by three different anthracycline and taxane containing regimens for early breast cancer. *Breast Cancer Res Treat.* 2009;115(2):335–342.

29. Zhou WB, Yin H, Liu XA, et al. Incidence of chemotherapy-induced amenorrhea associated with epirubicin, docetaxel and navelbine in younger breast cancer patients. *BMC Cancer.* 2010;10:281.

30. Lee S, Kil WJ, Chun M, et al. Chemotherapy-related amenorrhea in premenopausal women with breast cancer. *Menopause.* 2009;16(1):98–103.

31. Di Cosimo S, Alimonti A, Ferretti G, et al. Incidence of chemotherapy-induced amenorrhea depending on the timing of treatment by menstrual cycle phase in women with early breast cancer. *Ann Oncol.* 2004;15(7):1065–1071.

32. Siegel R, Naishadham D, Jemal A. Cancer statistics, 2012. *CA Cancer J Clin.* 2012;62(1):10–29.

33. Rodriguez-Wallberg KA, Oktay K. Options on fertility preservation in female cancer patients. *Cancer Treat Rev.* 2012;38(5):354–361.

34. Fornier MN, Modi S, Panageas KS, Norton L, Hudis C. Incidence of chemotherapy-induced, long-term amenorrhea in patients with breast carcinoma age 40 years and younger after adjuvant anthracycline and taxane. *Cancer.* 2005;104(8):1575–1579.

35. Gerber B, von Minckwitz G, Stehle H, German Breast Group Investigators, et al. Effect of luteinizing hormone–releasing hormone agonist on ovarian function after modern adjuvant breast cancer chemotherapy: the GBG 37 ZORO study. *J Clin Oncol.* 2011;29(17):2334–2341.

36. Berliere M, Dalenc F, Malingret N, et al. Incidence of reversible amenorrhea in women with breast cancer undergoing adjuvant anthracycline-based chemotherapy with or without docetaxel. *BMC Cancer.* 2008;8:56.

37. Sukumvanich P, Case LD, Van Zee K, et al. Incidence and time course of bleeding after long-term amenorrhea after breast cancer treatment: a prospective study. *Cancer.* 2010;116(13):3102–3111.

38. Storm HH, Klint A, Tryggvadottir L, et al. Trends in the survival of patients diagnosed with malignant neoplasms of lymphoid, haematopoietic, and related tissue in the Nordic countries 1964-2003 followed up to the end of 2006. *Acta Oncol.* 2010;49(5):694–712.

39. Hodgson DC, Pintilie M, Gitterman L, et al. Fertility among female hodgkin lymphoma survivors attempting pregnancy following ABVD chemotherapy. *Hematol Oncol.* 2007;25(1):11–15.

40. De Bruin ML, Huisbrink J, Hauptmann M, et al. Treatment-related risk factors for premature menopause following Hodgkin lymphoma. *Blood.* 2008;111(1):101–108.

41. Bedoschi G, Oktay K. Current approach to fertility preservation by embryo cryopreservation. *Fertil Steril.* 2013;99(6):1496–1502.

42. Whitehead E, Shalet SM, Blackledge G, Todd I, Crowther D, Beardwell CG. The effect of combination chemotherapy on ovarian function in women treated for Hodgkin's disease. *Cancer.* 1983;52(6):988–993.

43. Elis A, Tevet A, Yerushalmi R, et al. Fertility status among women treated for aggressive non-Hodgkin's lymphoma. *Leuk Lymphoma.* 2006;47(4):623–627.

44. Dann EJ, Epelbaum R, Avivi I, et al. Fertility and ovarian function are preserved in women treated with an intensified regimen of cyclophosphamide, adriamycin, vincristine and prednisone (Mega-CHOP) for non-Hodgkin lymphoma. *Hum Reprod.* 2005;20(8):2247–2249.

45. Belson M, Kingsley B, Holmes A. Risk factors for acute leukemia in children: a review. *Environ Health Perspect.* 2007;115(1):138–145.

46. Kreuser ED, Hetzel WD, Heit W, et al. Reproductive and endocrine gonadal functions in adults following multidrug chemotherapy for acute lymphoblastic or undifferentiated leukemia. *J Clin Oncol.* 1988;6(4):588–595.

47. Spinelli S, Chiodi S, Bacigalupo A, et al. Ovarian recovery after total body irradiation and allogeneic bone marrow transplantation: long-term follow up of 79 females. *Bone Marrow Transplant.* 1994;14(3):373–380.

48. Del Mastro L, Boni L, Michelotti A, et al. Effect of the gonadotropin releasing hormone analogue triptorelin on the occurrence of chemotherapy-induced early menopause in premenopausal women with breast cancer: a randomized trial. *JAMA.* 2011;306(3):269–276.

49. Munster PN, Moore AP, Ismail-Khan R, et al. Randomized trial using gonadotropin-releasing hormone agonist triptorelin for the preservation of ovarian function during (neo) adjuvant chemotherapy for breast cancer. *J Clin Oncol.* 2012;30(5):533–538.

50. Turner NH, Partridge A, Sanna G, Di Leo A, Biganzoli L. Utility of gonadotropin-releasing hormone agonists for fertility preservation in young breast cancer patients: the benefit remains uncertain. *Ann Oncol.* 2013;24(9):2224–2235.

51. Bedoschi G, Turan V, Oktay K. Utility of GnRH-agonists for fertility preservation in women with operable breast cancer: is it protective?. *Curr Breast Cancer Rep.* 2013;5(4):302–308.

52. Letourneau JM, Ebbel E, Katz P, et al. The prevalence of self-reported reproductive impairment in young female cancer survivors throuhgout California. *Fertil Steril.* 2010;94(4):510.

53. Meirow D, Dor J, Kaufman B, et al. Cortical fibrosis and blood-vessels damage in human ovaries exposed to chemotherapy. Potential mechanisms of ovarian injury. *Hum Reprod.* 2007;22(6):1626–1633.

54. Titus S, Li F, Stobezki R, et al. Impairment of *BRCA1*-related DNA double-strand break repair leads to ovarian aging in mice and humans. *Sci Transl Med.* 2013;5(172):172ra21

55. Kalich-Philosoph L, Roness H, Carmely A, et al. Cyclophosphamide triggers follicle activation and "burnout"; AS101 prevents follicle loss and preserves fertility. *Sci Transl Med.* 2013;5(185):185ra62

56. Meirow D, Epstein M, Lewis H, Nugent D, Gosden RG. Administration of cyclophosphamide at different stages of follicular maturation in mice: effects on reproductive performance and fetal malformations. *Hum Reprod.* 2001;16(4):632–637.

57. Kujjo LL, Chang EA, Pereira RJ, et al. Chemotherapy-induced late transgenerational effects in mice. *PLoS One.* 2011;6(3):e17877.

58. Meirow D, Lewis H, Nugent D, Epstein M. Subclinical depletion of primordial follicular reserve in mice treated with cyclophosphamide: clinical importance and proposed accurate investigative tool. *Hum Reprod.* 1999;14(7):1903–1907.

59. Oktay K, Schenken RS, Nelson JF. Proliferating cell nuclear antigen marks the initiation of follicular growth in the rat. *Biol Reprod.* 1995;53:295–301.

60. Raz A, Fisch B, Okon E, et al. Possible direct cytotoxicity effects of cyclophosphamide on cultured human follicles: an electron microscopy study. *J Assist Reprod Genet.* 2002;19 (10):500–506.

61. Morgan S, Lopes F, Gourley C, Anderson RA, Spears N. Cisplatin and doxorubicin induce distinct mechanisms of ovarian follicle loss; imatinib provides selective protection only against cisplatin. *PLoS One.* 2013;8(7):e70117.

62. Marcello MF, Nuciforo G, Romeo R, et al. Structural and ultrastructural study of the ovary in childhood leukemia after successful treatment. *Cancer.* 1990;66(10):2099–2104.

63. Hancke K, Strauch O, Kissel C, Gobel H, Schafer W, Denschlag D. Sphingosine 1-phosphate protects ovaries from chemotherapy-induced damage in vivo. *Fertil Steril.* 2007;87(1):172–177.

64. Paris F, Perez GI, Fuks Z, et al. Sphingosine 1-phosphate preserves fertility in irradiated female mice without propagating genomic damage in offspring. *Nat Med.* 2002;8(9):901–902.

65. Li F, Turan V, Lierman S, Cuvelier C, De Sutter P, Oktay K. Sphingosine-1-phosphate prevents chemotherapy-induced human primordial follicle death. *Hum Reprod.* 2014;29(1):107–113.

66. Titus S, Stobezki R, Turan V, Halicka D, Sutter PD, Oktay K. Sphingosine-1-Phosphate, A protector against chemotherapy-induced ovarian follicle apoptosis, does not diminish the effectiveness of chemotherapy. *Fertil Steril.* 2014;105(3):e39.

67. Oktay K, Kim JY, Barad D, Babayev SN. Association of *BRCA1* mutations with occult primary ovarian insufficiency: a possible explanation for the link between infertility and breast/ovarian cancer risks. *J Clin Oncol.* 2010;28(2):240–244.

68. Ford D, Easton DF, Peto J. Estimates of the gene frequency of *BRCA1* and its contribution to breast and ovarian cancer incidence. *Am J Hum Genet.* 1995;57(6):1457–1462.

69. Santoro N. *BRCA* mutations and fertility: do not push the envelope!. *Fertil Steril.* 2013;99 (6):1560.

70. Rzepka-Gorska I, Tarnowski B, Chudecka-Gøaz A, Górski B, Zielińska D, Toøoczko-Grabarek A. Premature menopause in patients with *BRCA1* gene mutation. *Breast Cancer Res Treat.* 2006;100(1):59–63.

71. Lin WT, Beattie M, Chen LM, et al. Comparison of age at natural menopause in *BRCA1/2* mutation carriers with a non-clinic-based sample of women in northern California. *Cancer.* 2013;119(9):1652–1659.

72. Finch A, Valentini A, Greenblatt E, et al. Frequency of premature menopause in women who carry a *BRCA1* or *BRCA2* mutation. *Fertil Steril.* 2013;99(6):1724–1728.

73. Wang ET, Pisarska MD, Bresee C, et al. *BRCA1* germline mutations may be associated with reduced ovarian reserve. *Fertil Steril.* 2014;. Available from: http://dx.doi.org/10.1016/ j.fertnstert.2014.08.014.

74. Oktay K, Moy F, Titus S, et al. Age-related decline in DNA repair function explains diminished ovarian reserve, earlier menopause, and possible oocyte vulnerability to chemotherapy in women with *BRCA* mutations. *J Clin Oncol.* 2014;32(10):1093–1094.

Laboratory Models

CHAPTER 5

In Vivo Models of Ovarian Toxicity

Elon C. Roti Roti[1], Sana M. Salih[2], and Mary B. Zelinski[3,4]

[1]Department of Medicine, University of Wisconsin, Madison, WI, USA; [2]Department of Obstetrics and Gynecology, University of Wisconsin, Madison, WI, USA; [3]Division of Reproductive & Developmental Sciences, Oregon National Primate Research Center, Oregon Health & Science University, Beaverton, OR, USA; [4]Department of Obstetrics & Gynecology, Oregon Health & Science University, Portland, OR, USA

5.1 INTRODUCTION

The success of cancer treatment creates a growing population of female cancer survivors who wrestle with long-term side effects of chemotherapy, including primary ovarian insufficiency (POI).[1-4] This growing survivorship presents researchers and clinicians with the challenging mandate to develop therapeutics that improve ovarian function post-chemotherapy. Continuing research is therefore devoted to developing non-invasive, drug-based ovarian shields and ovarian-regenerative technology.[5-14] An ideal ovoprotective drug would be easily administered in conjunction with traditional chemotherapy, and preserve both long-term endocrine function and fertility by preventing chemotherapy toxicity to the ovary. Developing ovoprotective drugs requires detailed understanding the mechanism(s) of chemotherapy ovarian insult in order to identify potential targets to achieve the ultimate balancing act: shield the healthy ovary from chemotherapy, but kill cancer. Key parameters include identifying the ovarian cell type(s) and follicles affected by each chemotherapy agent, and temporal resolution of primary insult and subsequent cell death. In vivo animal models are required to understand systemic effects of chemotherapy as they provide native cellular interactions, the dynamic intrafollicular communication and allow assessment of fertility and the health of subsequent generations.

5.2 RODENT MODELS OF CHEMOTHERAPY-INDUCED FOLLICULAR DEPLETION AND OVOPROTECTION

The most common in vivo model for ovotoxicity is the mouse. Rodents provide short lifespans with rapid progression from birth through

Cancer Treatment and the Ovary. DOI: http://dx.doi.org/10.1016/B978-0-12-801591-9.00005-9

reproductive maturity (weeks) and short intergenerational times (months), allowing comprehensive studies from acute insult through the birth of multiple generations. While not currently standard in ovotoxicity studies, transgenic mouse cancer models of leukaemia, breast cancer and sarcomas[15-17] will be key in determining whether chemotherapy remains effective when administered with ovoprotective agents by providing simultaneous assessment of cancer remission rates and ovoprotection. Despite these advantages, large differences remain compared to human reproductive physiology; rodents are multi-ovulatory and typically do not experience natural menopause. Rodents are more sensitive to chemotherapy than humans, limiting cumulative chemotherapy doses and impeding recapitulation of multi-dose chemotherapy regimens. Proof-of-concept oncofertility rodent studies therefore require confirmation in non-human primates more closely related to humans.

5.2.1 Chemotherapy-Induced Follicular Depletion

Foundational chemotherapy studies measured ovotoxicity as follicle loss, leading to three models describing follicular depletion. In the first model, chemotherapy directly destroys primordial follicles, depleting the ovarian reserve.[18-20] For example, cisplatin depletes primordial and primary follicles in mice treated at postnatal day (PND)5 (i.e., paediatric mice).[21,22] Secondary follicle counts are unchanged at PND9,[21,22] however, demonstrating relative resistance.

In the second model, chemotherapy induces apoptosis only in follicles containing actively dividing granulosa cells. Doxorubicin (DXR) induces widespread apoptosis in granulosa cells of growing follicles within 12 hours, but primordial follicles do not exhibit apoptosis until 48 hours post-treatment, and then only in small numbers consistent with the primordial follicle recruitment rate in mice.[23,24]

The third, not mutually exclusive, "burnout" model postulates that cyclophosphamide first depletes growing follicles, which promotes primordial follicle recruitment, explaining the observed time-dependent follicular depletion followed by restoration of growing follicle numbers.[25,26] This model predicts that consecutive chemotherapy treatments compound depletion as primordial follicles prematurely activated by the first round of chemotherapy become the growing follicle pool destroyed by the following chemotherapy dose. Future studies tagging chemotherapy-injured follicles to quantify growth and atresia *in vivo* may allow clear delineation of ovarian remodelling.

5.2.2 Cell-Type Specificity of Chemotherapy Toxicity

Early studies of chemotherapy toxicity placed significant focus on oocytes, including cisplatin destruction of primordial follicle oocytes.[22] Distinct chemotherapy agents exhibit differential toxicity profiles, however, with respect to which cell and follicle types are the primary targets for insult and demise. Stromal and thecal cells in direct contact with the blood supply are the first cells to exhibit DXR-induced DNA damage, followed by subsequent damage to follicular cells (Figure 5.1).[23,27] Identifying the cells that receive the first insult for each chemotherapy agent is an important step toward designing protective approaches that prevent ovotoxicity.

5.2.3 Ovarian Toxicity Lessons from Ovoprotection Studies

Mechanistic details of how chemotherapy causes ovarian damage continue to expand through studies testing ovoprotective agents. Here we highlight key considerations for rodent models, including animal age at treatment, drug preparation methods and drug dose, alongside the need to develop robust chemoprotective agents that provide protection across a range of chemotherapy doses.

The importance of drug preparation was revealed by conflicting results in cisplatin-induced ovarian damage in studies by Gonfloni and Kerr.[21,22] A thorough and mechanistic study by Gonfloni *et al.* demonstrating that imatinib shields the ovary from cisplatin damage showed an initial 40−50% loss of fertility and pups per litter following cisplatin.[21] In replicating that study, however, Kerr *et al.* observed almost complete loss of both.[22] Given identical calculated drug doses, mouse treatment and mating age, and that the Gonfloni group replicated their own results,[28] differences in cisplatin formulations may account for higher level of toxicity and the resultant lack of imatinib protection observed by Kerr *et al.*[22]

Comparing ovotoxicity across studies can be challenging when different chemotherapy doses are used. Cyclophosphamide at 50 mg/kg causes 60% primordial follicle loss in mice treated at 5−6 weeks of age.[29,30] Higher cyclophosphamide doses (120 mg/kg) combined with busulfan (12 mg/kg) cause a striking >95% reduction in primordial follicle counts.[31,32] While greater toxicity with higher dose is not surprising, these studies demonstrate challenges in cross-study

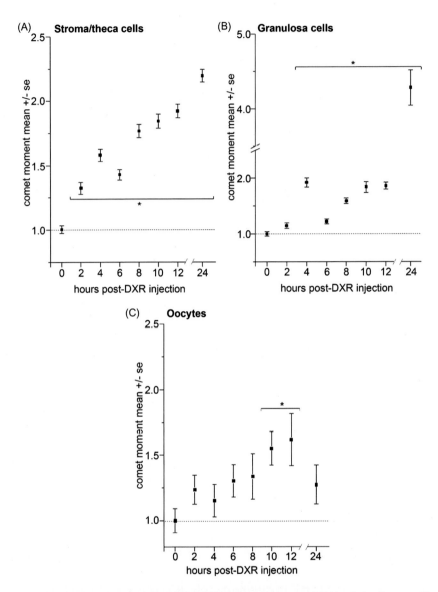

Figure 5.1 Temporal accumulation of doxorubicin (DXR)-induced double-strand DNA breaks is cell-type specific. Summary data quantify double-stranded DNA damage as the comet moment utilizing the comet assay. Panels summarize DNA damage in stromal/thecal cells (A), granulosa cells (B), and oocytes (C) as hours post-DXR injection plotted against comet moment. n = 3 animals/group/time point/replicate, >100 cells/point, 4 replicates total. *p < 0.05, one-way analysis of variance. Adapted from[23].

comparisons and effective ovoprotective agents will need to shield across a range of chemotherapy doses.

Similar to cancer patients, age of mice at time of treatment alters the characteristics of chemotherapy ovotoxicity. The burnout model explains dynamics of initial secondary follicle depletion and restoration in PND5 mice treated with 150 mg/kg cyclophosphamide.[25] In contrast, primordial follicles from mice treated at 4.5 weeks of age with 50 mg/kg cyclophosphamide were depleted *prior* to loss of growing follicles.[30] While the chemotherapy dose was different, altered follicular dynamics are likely age-dependent. Similarly, 4-week-old (adolescent) mice pretreated with granulocyte colony stimulating factor as an ovoprotective agent exhibited >80% follicle loss following cyclophosphamide/busulfan, but breeding mice only showed a 30% reduction in litter size.[33] These discrepancies may be due to the treatment of 8-week-old (reproductively mature) mice for fertility assessment.[33] As the field moves forward, systematic assessment of age-related chemotoxicity and efficacy of ovoprotective agents will aid translation of fertility interventions to cancer patients.

5.2.4 Rodent Studies of Ovarian Chemotherapy Toxicity Compared to Data from Clinical Studies

Chemotherapy agents have been classified as differentially ovotoxic based on retrospective analysis of female cancer survivors who received multi-drug treatments. Alkylating agents like cyclophosphamide are considered the most detrimentally ovotoxic. Platinum drugs, including cisplatin, are classified as moderately ovotoxic, and anthracyclines like DXR, mildly ovotoxic.[34] Rodent studies of cisplatin and cyclophosphamide confirm the human ovotoxicity, but mice receiving a single human equivalent dose of DXR exhibit a 50% decrease in ovarian mass and decline in fertility rates much greater than "mildly ovotoxic" would predict. It is unclear whether the disparities are due to species differences or the fact that infertility post-chemotherapy is underestimated, given that it requires self-reporting from patients, and that follow-up is usually for a relatively short period of time. In addition, clinical ovotoxicity has thus far only been scored according to infertility, not accounting for additional negative ramifications in females who do achieve pregnancy, including delivery complications and low birth weights. Sibling studies conclude that while alkylating agents (considered highly ovotoxic) do not affect birth weight, non-alkylating agents, particularly DXR, carry high risk for low birth-weight babies in subsequent pregnancies.[35,36] These data highlight the need for a comprehensive classification of chemotherapy ovotoxicity.

Cancer patients receive chemotherapy cocktails, but most animal models have assessed ovotoxicity of each chemotherapy agent in isolation. Advancing toward the goal of cross-chemotherapy ovoprotection, tamoxifen shields the ovary from both cyclophosphamide (*in vivo*) and doxorubicin (*in vitro*).[11] This study provides an important step toward animal models of combined chemotherapy regimens.

Another critical question in predicting POI following chemotherapy is determining how follicle counts predict POI and developing *in vivo* follicle assessment tools. Mice treated with cyclophosphamide exhibit a linear relationship between dose and primordial follicle depletion but, surprisingly, depleting 60% of primordial follicles does not change ovulation, fertilization, or pregnancy rates.[29] Certainly follicle depletion reduces the fertile window, but in women, removing one ovary does not dramatically change the age of menopause.[37] The processes allowing one ovary to compensate for the loss of the contralateral ovary are unknown, and may reveal novel ways to prolong fertility despite follicular depletion. Magnetic resonance imaging (MRI) has revealed DXR-induced ovarian remodelling in mice as a decrease in ovarian dimensions through 1-month post-DXR treatment.[24] Future studies correlating follicle counts with ovarian changes measured by MRI may provide a non-invasive measure of ovarian remodelling and POI assessment post-chemotherapy in patients.

5.2.5 Transgenerational Toxicity of Chemotherapy

Current dogma suggests that women who become pregnant post-cancer are at no increased risk for offspring with congenital defects. Data from mice, however, demonstrate that while the first generation post-chemotherapy manifests no gross defects, chromosomal abnormalities appear in later generations.[38] In a transplant model that distinguishes between effects on the ovary itself and general physiological decline of the dam, ovaries from DXR-treated mice were transplanted into naïve mice, which were subsequently mated. Starting at generation (G)4, DXR offspring exhibited increased neonatal death and physical abnormalities due to deletions on chromosome 10. Dams receiving DXR ovaries suffered frequent dystocia not observed in mice receiving control ovaries, and died due to delivery complications in G4–G6. Similarly, oocytes retrieved from cyclophosphamide-treated mice exhibited abnormalities that led to reduced fertilization and embryonic development, and increased aneuploidy.[39] These data suggest transgenerational effects of chemotherapy may be underestimated in human cancer survivors, as society has not reached five generations post-chemotherapy (as these agents were first developed in the 1960s) and demonstrate that successful ovoprotective agents will need to *prevent* chemotherapy-induced genotoxicity, diminishing risk for propagating chemotherapy-induced genetic abnormalities.

Mice provide the ideal model for initiating studies to determine whether mechanistic ovoprotective agents like dexrazoxane and bortezomib can prevent chemotherapy-induced genotoxicity[8,9] and limit risk for future generations. Demonstrating the strengths of an *in vivo* model, including thorough analysis of acute insult to all ovarian cell types combined with long-term fertility assessment, a recent study by the authors showed that bortezomib shields granulosa, stromal/thecal cells and oocytes of adolescent mice (4 weeks of age) from DXR-induced DNA damage in a temporal fashion, preserving the growing follicle pool.[8] While bortezomib did not alter initial DXR-induced loss, litter size recovered over time compared to DXR alone (Figure 5.2). The temporal data provided a thorough assessment of fertility across the reproductive lifespan of the mice and revealed that initial fertility loss can be recovered. Future studies will determine whether mechanistic ovoprotective agents fulfil the potential of diminishing the transgenerational consequences of chemotherapy.

5.3 PRIMATE MODELS OF CHEMOTHERAPY-INDUCED OVARIAN TOXICITY AND OVOPROTECTION

Rhesus macaques recapitulate the human reproductive system with the greatest fidelity and therefore represent a preferred animal model for validating ovoprotective agents. Macaque monkeys ovulate one oocyte approximately every 28 days, and menstruate ~14 days after ovulation. They reach puberty at 5–6 years of age, often achieve conception during their first ovarian cycle and have an average 176 day gestation period.[40] However, reproductive studies require years to follow from treatment through breeding and subsequent generations. A complete cadre of assisted reproductive technologies (ARTs) have been implemented in rhesus monkeys, including ovarian stimulation, *in vitro* oocyte maturation,

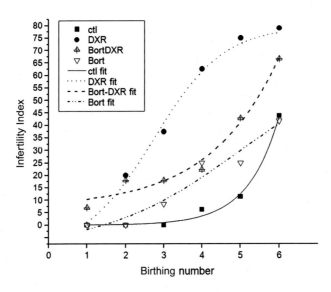

Figure 5.2 **Bort pretreatment improves the infertility index following doxorubicin (DXR) treatment across six rounds of continuous breeding.** *Plot represents the "infertility index" plotted as the percentage of surviving animals from each treatment group that fail to achieve the next birth as a function of birth (litter) number. Symbols correspond to control, DXR, Bort-DXR and Bort treatment groups as indicated. DXR points in the linear range were statistically different from Bort-DXR, Bort, and control (p < 0.05, one-way analysis of variance).* Adapted from[8].

in vitro fertilization, intracytoplasmic sperm injection, embryo culture and embryo transfer.[41,42] These techniques would easily allow acute examination of follicular and oocyte function, as well as embryonic effects, pre- and post-chemotherapy with an ovoprotective agent. Mechanisms of chemotherapy and ovoprotective agents should be studied in ovaries from animals of varying ages, which is critical given that ovarian reserve and response to anti-cancer therapies are age-dependent.

Marmoset monkeys may provide another model for ovarian toxicity, offering a shorter intergenerational time than macaques. Marmosets form stable families and ovulate 2−3 oocytes every 28 days. Adolescent female marmosets at 11−13 months are developmentally equivalent to adolescent human girls at puberty. Small in size, marmosets achieve sexual maturity earlier than macaques, with first conceptions occurring around 18 months and first births at 24 months of age.[43,44] These short intergenerational intervals provide a translatable *in vivo* primate model for examining fertility and transgenerational endpoints following ovarian chemotoxicity and ovoprotection.[45,46] ARTs in marmosets are not well developed, however. *In vitro* fertilization has been a particular challenge due to inadequate gonadotropin response, lack of ovarian ultrasound imaging during gonadotropin stimulation, and poor fertilization rates and embryo quality compared to rhesus macaques and women.[47,48]

Non-human primates have been used in longitudinal chemotherapy toxicity studies lasting over 30 years.[49] Despite similarities between the non-human primate and human ovary, challenges remain in using these models for radiotherapy and chemotherapy ovarian toxicity. Non-human primates have lower tolerability and survivability to chemotherapy than humans; chemotherapy doses routinely given to children can cause a myriad of complications in non-human primates. Standard human equivalent doses of DXR cause morbidity in macaques,[50] such that relatively low doses at 1–2 mg/kg are routinely used, with a maximum of 5.83 mg/kg dose, which is <50% human equivalent.[49–57] Similarly, rhesus monkeys treated with a single dose (4 to 12 mg/kg) of the alkylating chemotherapy, busulfan, manifested significant dose-dependent testicular insufficiency.[58] Busulfan is used in humans at a total dose of 16 mg/kg (4 mg/kg daily for 4 days) for conditioning prior to bone marrow transplant.[59] The 16 mg/kg total human dose is equivalent to around 50 mg/kg in the rhesus (based on Km = 37 for a 60 mg adult with a body surface area of 1.2 m^2 and rhesus Km = 12).[60] Rhesus monkeys showed increased mortality due to severe bone marrow suppression at the lower 8 mg/kg dose.[58,61]

Although non-human primates cannot tolerate precise human equivalent doses, developing ovoprotective agents requires defining the tolerable maximal dose that elicits ovarian toxicity mimicking the response in women who receive systemic chemotherapy. To maintain animal health, low doses of systemic chemotherapy must consistently be used in conjunction with vigorous prophylactic supportive care, e.g., apheresis.[58] Administration via arterial vessels localized near the ovary[62] or intraovarian delivery systems[5] may circumvent detrimental systemic side effects of chemotherapeutic agents while maintaining ovarian toxicity and providing routes for simultaneous treatment with ovoprotectants. Although malignant lymphomas and leukaemias have been described in macaques and marmosets for many decades, the majority do not arise spontaneously, but are reported in animals with simian immunodeficiency virus-related immune suppression and associated with viral infections.[63] Burkitt's type and T-cell lymphomas can be virally induced (i.e., via lymphocryptovirus and simian T lymphotropic virus, respectively), but viral associated leukaemias are very rare.[63] Thus, non-human primates are limited as naturally occurring cancer models, but haematological disease can be induced with oncogenic viruses.[64] Despite these challenges, establishing non-human primate models to pre-clinically evaluate chemotherapy-induced ovarian toxicity *in vivo* is a priority if development and delivery of ovoprotective agents shown to be promising in rodents are to become a clinical reality.

One proof-of-concept study for ovarian protection based on previous mechanistic studies in rodents[65] has been performed in female rhesus macaques, wherein intraovarian administration of the anti-apoptotic agent, sphingosine-1-phosphate (S1P) or its long-acting analogue, FTY720, preceding ovarian X-irradiation at a dose causing ovarian failure (15 Gy) sustained a cohort of preantral follicles.[5] Ovarian cyclicity resumed in treated animals, and fertility was restored resulting in live offspring devoid of chromosomal abnormalities.[5] The protective effect of S1P against ovarian failure caused by chemotherapeutic agents was corroborated by subsequent *in vivo* studies in rodents exposed to the alykylating agent dacarbazine,[66] as well as human ovarian cortex treated with cyclophosphamide or DXR.[10] The precise cellular targets (vascular, somatic, oocyte) and protective mechanism of S1P in the primate ovary are unknown. A disadvantage of S1P is that is must be delivered locally within the ovary and not systemically where it could potentially protect malignant cells from chemotherapy. Nonetheless, these studies demonstrate the utility of the non-human primate model for *in vivo* studies on prevention of ovarian damage from clinical cancer therapies by ovoprotectants that are potentially effective in women.

5.4 CONCLUSION

In vivo studies in rodents have provided critical information on the sites and mechanisms whereby various chemotherapy agents cause ovarian toxicity. While mechanistic studies of ovarian chemotherapy toxicity performed *in vitro* are valuable, they must ultimately be performed *in vivo* under chronic (months vs. one week) exposure, preferably in non-human primate models that best mimic ovarian function in women. *In vivo* studies are notably complex in that animal morbidity must be closely monitored and controlled. To better serve the future fertility needs of female paediatric, adolescent and young adult populations who are primarily survivors of leukaemias and lymphomas, more research on cyclophosphamide is needed in a non-human primate model in addition to current research on cisplatin, DXR and busulfan. Emerging *in vivo* rodent studies reveal chemotherapeutic agents may target multiple ovarian cell types acutely, allowing some recovery of ovarian function, or chronically, thereby preventing long-term fertility preservation. Delineating the temporal events and cellular mechanisms leading to ovarian damage within individual compartments will be necessary to establish first-line and downstream protective measures that can maintain ovarian function and future fertility without interfering with cancer therapy. Pharmacological approaches to preserve female fertility and reproductive health based on specific mechanisms of ovarian toxicity constitute one important method for ovarian chemoprotection. Future research should also, however, emphasize the design of novel chemotherapeutic agents that are not toxic to the ovary. Either strategy has the potential to totally transform current clinical oncofertility practices for younger female cancer patients.

REFERENCES

1. Woodruff TK. The oncofertility consortium – addressing fertility in young people with cancer. *Nat Rev Clin Oncol*. 2010;7:466–475.

2. Hewitt M, Breen N, Devesa S. Cancer prevalence and survivorship issues: Analyses of the 1992 National Health Interview Survey. *J Natl Cancer Inst*. 1999;91:1480–1486.

3. Oktay K, Oktem O. Fertility preservation medicine: a new field in the care of young cancer survivors. *Pediatr Blood Cancer*. 2009;53:267–273.

4. Oktay K, Oktem O, Reh A, Vahdat L. Measuring the impact of chemotherapy on fertility in women with breast cancer. *J Clin Oncol*. 2006;24:4044–4046.

5. Zelinski MB, Murphy MK, Lawson MS, Jurisicova A, Pau KYF, Toscano NP, et al. *In vivo* delivery of FTY720 prevents radiation-induced ovarian failure and infertility in adult female nonhuman primates. *Fertil Steril*. 2011;95:1440-U289.

6. Kim SY, Cordeiro MH, Serna VA, Ebbert K, Butler LM, Sinha S, et al. Rescue of platinum-damaged oocytes from programmed cell death through inactivation of the p53 family signaling network. *Cell Death Differ*. 2013;20:987–997.

7. Roness H, Kalich-Philosoph L, Carmely A, Fishel-Bartal M, Ligumsky H, Paglin S, et al. Cyclophosphamide triggers follicle activation causing ovarian reserve 'burn out'; AS101 prevents follicle loss and preserves fertility. *Hum Reprod*. 2013;28:46–47.

8. Roti Roti EC, Ringelstetter AK, Kropp JK, Abbott DH, Salih SM. Bortezomib prevents acute doxorubicin ovarian insult and follicle demise, improving the fertility window and pup birth weight in mice. *PLoS One*. 2014;9(9):e108174.

9. Roti Roti EC, Salih SM. Dexrazoxane ameliorates doxorubicin-induced injury in mouse ovarian cells. *Biol Reprod*. 2012;86:1–11.

10. Li F, Turan V, Lierman S, Cuvelier C, De Sutter P, Oktay K. Sphingosine-1-phosphate prevents chemotherapy-induced human primordial follicle death. *Hum Reprod*. 2014;29:107–113.

11. Ting AY, Petroff BK. Tamoxifen decreases ovarian follicular loss from experimental toxicant DMBA and chemotherapy agents cyclophosphamide and doxorubicin in the rat. *J Assist Reprod Genet*. 2010;27:591–597.

12. Urruticoechea A, Arnedos M, Walsh G, Dowsett M, Smith IE. Ovarian protection with goserelin during adjuvant chemotherapy for pre-menopausal women with early breast cancer (EBC). *Breast Cancer Res Treat*. 2008;110:411−416.

13. Parte S, Bhartiya D, Manjramkar DD, Chauhan A, Joshi A. Stimulation of ovarian stem cells by follicle stimulating hormone and basic fibroblast growth factor during cortical tissue culture. *J Ovarian Res*. 2013;6(1):20.

14. Liu T, Huang YY, Guo LH, Cheng WW, Zou G. CD44+/CD105+ Human amniotic fluid mesenchymal stem cells survive and proliferate in the ovary long-term in a mouse model of chemotherapy-induced premature ovarian failure. *Int J Med Sci*. 2012;9:592−602.

15. Ahn RW, Barrett SL, Raja MR, Jozefik JK, Spaho L, Chen H, et al. Nano-encapsulation of arsenic trioxide enhances efficacy against murine lymphoma model while minimizing its impact on ovarian reserve *in vitro* and *in vivo*. *Plos One*. 2013;8(3):e58491.

16. Hollern DP, Andrechek ER. A genomic analysis of mouse models of breast cancer reveals molecular features of mouse models and relationships to human breast cancer. *Breast Cancer Res*. 2014;16(3):R59.

17. Post SM. Mouse models of sarcomas: critical tools in our understanding of the pathobiology. *Clin Sarcoma Res*. 2012;2(1):20.

18. Borovskaya TG, Goldberg VE, Fomina TI, Pakhomova AV, Kseneva SI, Poluektova ME, et al. Morphological and functional state of rat ovaries in early and late periods after administration of platinum cytostatics. *Bull Exp Biol Med*. 2004;137:331−335.

19. Yeh J, Kim B, Liang YJ, Peresie J. Müllerian inhibiting substance as a novel biomarker of cisplatin-induced ovarian damage. *Biochem Biophys Res Commun*. 2006;348:337−344.

20. Yucebilgin MS, Terek MC, Ozsaran A, Akercan F, Zekioglu O, Isik E, et al. Effect of chemotherapy on primordial follicular reserve of rat: an animal model of premature ovarian failure and infertility. *Aust N Z J Obstet Gynaecol*. 2004;44:6−9.

21. Gonfloni S, Di Tella L, Caldarola S, Cannata SM, Klinger FG, Di Bartolomeo C, et al. Inhibition of the c-Abl-TAp63 pathway protects mouse oocytes from chemotherapy-induced death. *Nat Med*. 2009;15:1179−1185.

22. Kerr JB, Hutt KJ, Cook M, Speed TP, Strasser A, Findlay JK, et al. Cisplatin-induced primordial follicle oocyte killing and loss of fertility are not prevented by imatinib. *Nat Med*. 2012;18:1170−1172.

23. Roti Roti EC, Leisman SK, Abbott DH, Salih SM. Acute doxorubicin insult in the mouse ovary is cell- and follicle-type dependent. *Plos One*. 2012;7(8):e42293.

24. Ben-Aharon I, Bar-Joseph H, Tzarfaty G, Kuchinsky L, Rizel S, Stemmer SM, et al. Doxorubicin-induced ovarian toxicity. *Reprod Biol Endocrinol*. 2010;8:20.

25. Kalich-Philosoph L, Roness H, Carmely A, Fishel-Bartal M, Ligumsky H, Paglin S, et al. Cyclophosphamide triggers follicle activation and "Burnout"; AS101 prevents follicle loss and preserves fertility. *Sci Transl Med*. 2013;5(185):185ra62.

26. Meirow D, Philosof-Kalich L, Carmely A, Bartal M, Roness H. Follicle "burn out": a novel mechanism of chemotherapy induced ovarian damage. *Fertil Steril*. 2010;94:S10−S.

27. Soleimani R, Heytens E, Darzynkiewicz Z, Oktay K. Mechanisms of chemotherapy-induced human ovarian aging: double strand DNA breaks and microvascular compromise. *Aging (US)*. 2011;3(8):782−793.

28. Maiani E, Di Bartolomeo C, Klinger FG, Cannata SM, Bernardini S, Chateauvieux S, et al. Cisplatin-induced primordial follicle oocyte killing and loss of fertility are not prevented by imatinib Reply. *Nat Med*. 2012;18:1172−1174.

29. Meirow D, Lewis H, Nugent D, Epstein M. Subclinical depletion of primordial follicular reserve in mice treated with cyclophosphamide: clinical importance and proposed accurate investigative tool. *Hum Reprod*. 1999;14:1903−1907.

30. Kamarzaman S, Shaban M, Rahman SA. The prophylactic effect of nighella sativa against cyclophosphamide in the ovarian follicles of matured adult mice: a preliminary study. *J Anim Plant Sci.* 2014;24:81−88.

31. Lee HJ, Selesniemi K, Niikura Y, Niikura T, Klein R, Dombkowski DM, et al. Bone marrow transplantation generates immature oocytes and rescues long-term fertility in a preclinical mouse model of chemotherapy-induced premature ovarian failure. *J Clin Oncol.* 2007;25:3198−3204.

32. Santiquet N, Vallieres L, Pothier F, Sirard MA, Robert C, Richard F. Transplanted bone marrow cells do not provide new oocytes but rescue fertility in female mice following treatment with chemotherapeutic agents. *Cell Reprogram.* 2012;14:123−129.

33. Skaznik-Wikiel ME, McGuire MM, Sukhwani M, Donohue J, Chu TJ, Krivak TC, et al. Granulocyte colony-stimulating factor with or without stem cell factor extends time to premature ovarian insufficiency in female mice treated with alkylating chemotherapy. *Fertil Steril.* 2013;99(7):2045−2054.e3.

34. Meirow D, Biederman H, Anderson RA, Wallace WHB. Toxicity of chemotherapy and radiation on female reproduction. *Clin Obstet Gynecol.* 2010;53:727−739.

35. Green DM, Sklar CA, Boice JD, Mulvihill JJ, Whitton JA, Stovall M, et al. Ovarian failure and reproductive outcomes after childhood cancer treatment: results from the childhood cancer survivor study. *J Clin Oncol.* 2009;27:2374−2381.

36. Bar J, Davidi O, Goshen Y, Hod M, Yaniv I, Hirsch R. Pregnancy outcome in women treated with doxorubicin for childhood cancer. *Am J Obstet Gynecol.* 2003;189:853−857.

37. Bjelland EK, Wilkosz P, Tanbo TG, Eskild A. Is unilateral oophorectomy associated with age at menopause? A population study (the HUNT2 Survey). *Hum Reprod (Oxford, England).* 2014;29:835−841.

38. Kujjo LL, Chang EA, Pereira RJG, Dhar S, Marrero-Rosado B, Sengupta S, et al. Chemotherapy-induced late transgenerational effects in mice. *Plos One.* 2011;6(3):e17877.

39. Barekati Z, Gourabi H, Valojerdi MR, Yazdi PE. Previous maternal chemotherapy by cyclophosphamide (Cp) causes numerical chromosome abnormalities in preimplantation mouse embryos. *Reprod Toxicol.* 2008;26:278−281.

40. Fujita S, Sugiura H, Mitsunaga F, Shimizu K. Hormone profiles and reproductive characteristics in wild female Japanese macaques (Macaca fuscata). *Am J Primatol.* 2004;64:367−375.

41. Stouffer RL, Zelinski-Wooten MB. Overriding follicle selection in controlled ovarian stimulation protocols: quality vs quantity. *Reprod Biol Endocrinol.* 2004;2:32.

42. Ouhibi N, Zelinski-Wooten MB, Thomson JA, Wolf DP. Assisted fertilization and nuclear transfer in nonhuman primates. *Assist Fert Nucl Transf Mammals.* 2001;253−284.

43. Saltzman W, Digby LJ, Abbott DH. Reproductive skew in female common marmosets: what can proximate mechanisms tell us about ultimate causes? *Proc Biol Sci.* 2009;276:389−399.

44. Abbott DH, Barnett DK, Colman RJ, Yamamoto ME, Schultz-Darken NJ. Aspects of common marmoset basic biology and life history important for biomedical research. *Comp Med.* 2003;53:339−350.

45. 't Hart BA, Abbott DH, Nakamura K, Fuchs E. The marmoset monkey: a multi-purpose preclinical and translational model of human biology and disease. *Drug Discov Today.* 2012;17(21−22):1160−1165.

46. Orsi A, Rees D, Andreini I, Venturella S, Cinelli S, Oberto G. Overview of the marmoset as a model in nonclinical development of pharmaceutical products. *Regul Toxicol Pharmacol.* 2011;59:19−27.

47. Wilton LJ, Marshall VS, Piercy EC, Moore HDM. *In vitro* fertilization and embryo development in the marmoset monkey (*Callithrix jacchus*). *J Reprod Fertil.* 1993;97:481−486.

48. Marshall VS, Browne MA, Knowles L, Golos TG, Thomson JA. Ovarian stimulation of marmoset monkeys (Callithrix jacchus) using recombinant human follicle stimulating hormone. *J Med Primatol.* 2003;32:57–66.

49. Thorgeirsson UP, Dalgard DW, Reeves J, Adamson RH. Tumor incidence in a chemical carcinogenesis study of nonhuman primates. *Regul Toxicol Pharmacol.* 1994;19:130–151.

50. Gralla EJ, Fleischman RW, Luthra YK, Stadnicki SW. Dosing schedule dependent toxicities of adriamycin in beagle dogs and rhesus-monkeys. *Toxicology.* 1979;13:263–273.

51. Takayama S, Thorgeirsson UP, Adamson RH. Chemical carcinogenesis studies in nonhuman primates. *Proc Jpn Acad Ser B Phys Biol Sci.* 2008;84:176–188.

52. Schoeffner DJ, Thorgeirsson UP. Susceptibility of nonhuman primates to carcinogens of human relevance. *In Vivo.* 2000;14:149–156.

53. Sieber SM, Correa P, Young DM, Dalgard DW, Adamson RH. Cardiotoxic and possible leukemogenic effects of adriamycin in nonhuman primates. *Pharmacology.* 1980;20:9–14.

54. Warren KE, McCully CM, Walsh TJ, Balis FM. Effect of fluconazole on the pharmacokinetics of doxorubicin in nonhuman primates. *Antimicrob Agents Chemother.* 2000;44:1100–1101.

55. Warren KE, Patel MC, McCully CM, Montuenga LM, Balis FM. Effect of P-glycoprotein modulation with cyclosporin A on cerebrospinal fluid penetration of doxorubicin in nonhuman primates. *Cancer Chemother Pharmacol.* 2000;45:207–212.

56. Bronson RT, Henderson IC, Fixler H. Ganglioneuropathy in rabbits and a rhesus monkey due to high cumulative doses of doxorubicin. *Cancer Treat Rep.* 1982;66:1349–1355.

57. Berg SL, Reid J, Godwin K, Murry DJ, Poplack DG, Balis FM, et al. Pharmacokinetics and cerebrospinal fluid penetration of daunorubicin, idarubicin, and their metabolites in the nonhuman primate model. *J Pediatr Hematol Oncol.* 1999;21:26–30.

58. Hermann BP, Sukhwani M, Lin CC, Sheng Y, Tomko J, Rodriguez M, et al. Characterization, cryopreservation, and ablation of spermatogonial stem cells in adult rhesus Macaques. *Stem Cells.* 2007;25:2330–2338.

59. Socie G, Clift RA, Blaise D, Devergie A, Ringden O, Martin PJ, et al. Busulfan plus cyclophosphamide compared with total-body irradiation plus cyclophosphamide before marrow transplantation for myeloid leukemia: long-term follow-up of 4 randomized studies. *Blood.* 2001;98:3569–3574.

60. Reagan-Shaw S, Nihal M, Ahmad N. Dose translation from animal to human studies revisited. *FASEB J.* 2008;22:659–661.

61. Hermann BP, Sukhwani M, Winkler F, Pascarella JN, Peters KA, Sheng Y, et al. Spermatogonial stem cell transplantation into rhesus testes regenerates spermatogenesis producing functional sperm. *Cell Stem Cell.* 2012;11:715–726.

62. Sargent EL, Baughman WL, Novy MJ, Stouffer RL. Intraluteal infusion of a prostaglandin synthesis inhibitor, sodium meclofenamate, causes premature luteolysis in rhesus-monkeys. *Endocrinology.* 1988;123:2261–2269.

63. Miller AD. Neoplasia and proliferative disorders of nonhuman primates. In: Abee CR, Mansfield K, Tardif S, et al. , eds. *Nonhuman Primates in Biomedical Research: Diseases.* 2nd ed. London, UK: Elsevier; 2012:343–345.

64. Bruce AG, Bielefeldt-Ohmann H, Barcy S, Bakke AM, Lewis P, Tsai C-C, et al. Macaque homologs of EBV and KSHV show uniquely different associations with simian AIDS-related lymphomas. *Plos Pathogens.* 2012;8(10):e1002962.

65. Tilly JL, Kolesnick RN. Sphingolipids, apoptosis, cancer treatments and the ovary: investigating a crime against female fertility. *Biochim Biophys Acta.* 2002;1585:135–138.

66. Hancke K, Strauch O, Kissel C, Gobel H, Schafer W, Denschlag D. Sphingosine 1-phosphate protects ovaries from chemotherapy-induced damage *in vivo. Fertil Steril.* 2007;87:172–177.

CHAPTER 6

In Vitro Models of Ovarian Toxicity

Stephanie Morgan and Norah Spears

Centre for Integrative Physiology, University of Edinburgh, Edinburgh, UK

6.1 INTRODUCTION

Investigating the precise effect of chemotherapy on female fertility is a key research aspect underpinning fertility-preservation work. Protecting the ovarian follicle pool will require detailed understanding of precisely how the different chemotherapy drugs damage the ovary: whether chemotherapy drugs directly kill oocytes or whether damage is first to surrounding somatic cells; if the drugs are equally toxic to all follicles or if specific stages of follicle development are particularly vulnerable to damage. The more information that is available, the more informed will be our development of clinical treatments to protect the ovarian reserve. So far, *in vitro* techniques have been used relatively little for research into effects of chemotherapy treatment and potential ovarian protection against damage, but work to date indicates that these are, in general, powerful tools for reproductive toxicology studies, and so also likely to be very useful here.[1]

6.2 WHY USE CULTURE SYSTEMS?

Tissue culture systems have been extensively used to study key mechanisms and processes in the ovary, including follicle recruitment, oocyte–somatic cell interactions and follicle–follicle interactions. There are a variety of culture systems available, short- and long-term, most covering specific stages of follicle development. The addition of chemotherapeutic drugs into such systems allows determination of the effect of these agents on the health of various components of the ovary. Depending on the aims of the study, the molecular mechanisms of cell and follicle death can be established, along with examination of the effect on individual ovarian cell types, specific follicle stages and

Cancer Treatment and the Ovary. DOI: http://dx.doi.org/10.1016/B978-0-12-801591-9.00006-0

elucidation of primary versus secondary effects. A major consideration when studying the ovary is that the follicle population within it is heterogeneous, with follicles in varying stages of development: in postpubertal females, this includes follicles from the primordial right through to the preovulatory stage. This complicates *in vivo* research into the effect of chemotherapy drugs, since it is difficult to ascertain the initial site of damage following chemotherapy exposure. *In vitro* models can span various, yet precise, stages of follicle development, allowing for analysis of specific cell and follicle types.

While *in vitro* models can maintain follicle–follicle interactions and the influence of stromal cells essential for physiological ovarian function, culture does, of course, involve isolating the ovary from its neuroendocrine inputs, which also impacts on follicle recruitment and growth. *In vitro* models are therefore useful for determining the direct effect of chemotherapeutic agents on the ovary, but do not allow examination of indirect ovarian effects, such as occurs following hypothalamic/pituitary damage during cranial irradiation.[2]

When using *in vitro* techniques to examine the effects of chemotherapeutic agents, there are several limitations that must be considered. Premenopausal women receive complicated treatment regimens often involving combinations of several different drugs, generally given repeatedly over many weeks, which can be difficult to replicate. It is also hard to extrapolate the amount of drug that reaches the ovary solely from known serum concentrations, making it challenging to determine clinically relevant dosages to use *in vitro*. Chemotherapeutic agents can be metabolized *in vivo*, often in sites away from the ovary, such as the liver or the kidney, with the metabolites potentially having differing levels of ovotoxicity. An example of this is cyclophosphamide, which is metabolized by hepatic enzymes into a range of active metabolites including phosphoramide mustard and acrolein.[3] Utilizing the exact combination of metabolites that the ovary could be exposed to is therefore difficult to simulate *in vitro*, as is the metabolism of a drug if the metabolites themselves are not readily commercially available. There may also be limited information on the full range of active metabolites. Metabolites may have varying half-lives, even existing only transiently, complicating an experiment that is trying to replicate clinical scenarios. The volatility of some chemotherapy agents/metabolites is another consideration, as these can produce "next well" effects, requiring different treatment groups to be physically separated during culture.[4]

Overall, whilst *in vitro* techniques are valuable, they cannot simply replace *in vivo* experiments. The clinical scenario will involve indirect effects and potential input from other organs. Thus, *in vivo* systems are required to determine secondary effects from other organs, including changes to the hypothalamic−pituitary−gonadal axis. Nonetheless, use of *in vitro* culture systems reduces the number of animal experimentations required and allows for screening of a large range of drugs and drug doses. In particular, high drug dosages can be investigated more easily *in vitro*, since the experiments are not complicated by deteriorations in animal health or subsequent mortality. Overall, therefore, *in vivo* and *in vitro* methods both have their advantages and disadvantages, and both should be used for thorough investigation of reproductive toxicology.[1]

6.3 WHAT CULTURE SYSTEMS ARE AVAILABLE?

At present, a variety of culture systems are available, using tissue from humans, non-human primates, various domestic species and rodents, and each supporting different stages of follicle development (Figure 6.1). Within the ovarian cortex, the vast majority of follicles are primordial and it is this follicle reserve that dictates the reproductive lifespan of a woman. Therefore, a culture method that can evaluate the potential direct effect of chemotherapy drugs on primordial and early-stage follicles is one of the most clinically relevant systems. The ultimate culture method will support development of oocytes from primordial follicles through to fully mature, fertilizable oocytes *in vitro*. This is a major goal for reproductive biology, particularly with regards to human fertility preservation, but which to date has been achieved only in the mouse.

Primordial follicles do not survive culture well when isolated from ovarian tissue, most likely due to their small size and to the removal of support from the surrounding stroma and dense extracellular matrix. *In vitro* systems need to support the growth and activation of these follicles, which can be achieved through culturing cortical fragments (e.g., primates and ruminants) or culturing whole neonatal ovaries (e.g., rodents). Fragmentation of ovaries is also an increasingly utilized culture technique, which in rodents appears to promote follicle growth.[5,6]

At present, development of germ cells from primordial follicles through to the production of live offspring has been achieved only in

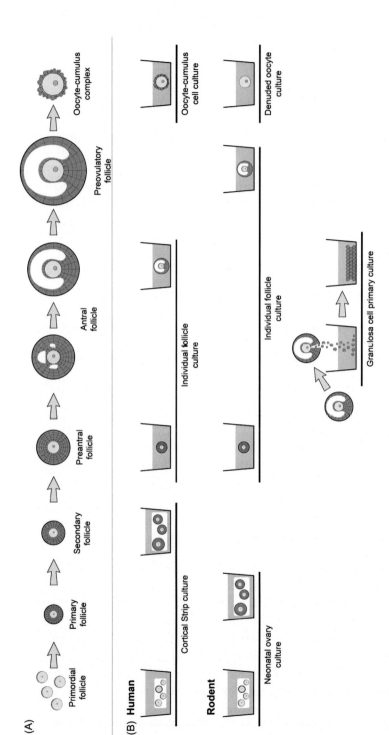

Figure 6.1 *Range of ovarian culture systems used to investigate chemotherapy toxicity.* (A) *Schematic of follicle development: follicles are activated from the resting pool of primordial follicles and progress through various stages of growth and development until they reach the preovulatory stage: at this point, ovulation occurs and the oocyte–cumulus complex is released.* (B) *Examples of human and rodent ovarian culture systems used for toxicity testing, spanning various stages of follicle development.*

the mouse.[7,8] In humans, whilst great progress has been made in this direction, culture systems are yet to be fully developed to the point where they can be utilized clinically.[9] Since offspring have not been obtained from starting material of primordial follicles outside of the mouse, it is, at this point, unclear for other species quite how closely physiological conditions have been replicated *in vitro*. Current experimental protocols aiming at long-term follicle development using ovaries from large mammals usually utilize one of two culture techniques. One approach is to use a multi-step culture system, which involves cortical strip culture to support primordial follicle activation followed by microdissection and individual follicle culture to the antral stage[10]: the final stage involves removing the oocyte–cumulus cells from the antral follicle and *in vitro* maturation of the oocyte followed by fertilization. Alternatively, immature follicles are cultured within a support matrix such as alginate hydrogels; to date, success has been achieved in culturing non-human primate individual secondary follicles contained in alginate to the antral stage.[11]

6.4 HUMAN AND OTHER PRIMATE MODEL STUDIES OF OVARIAN CHEMOTHERAPY TOXICITY

Human ovarian tissue for culture is usually obtained from one of three sources: fetal ovaries following termination of pregnancy; biopsies from women undergoing caesarean section/gynaecological surgery; or from fertility-preservation procedures. Depending on the age of the tissue at the time of collection, fetal ovaries may have germ cells that have not yet formed follicles. Furthermore, the immature tissue could potentially respond differently to insult than a mature ovary. Thus, fetal tissue is not the most appropriate model in which to investigate chemotherapy-induced toxicity in adults, although it is of course necessary to study effects of *in utero* exposure.[12] Tissue supply from women of the most clinically relevant age group is often limited, with the most desirable source being biopsies from young women, likely to have ovaries containing a relatively high density of follicles. As patient age increases, total follicle number declines, making it less likely that a biopsy will contain follicles for analysis. In addition, ovarian biopsies from older women usually have a high heterogeneity of follicle distribution, making it difficult to predict whether a specific piece of tissue contains follicles and making it hard for follicles to be evenly divided across treatment groups (if even present at all). One potential way around this is the use of vital dyes such as neutral red to identify the presence of follicles within slices of cortex before culture.[13]

Due to the limited quantity of human tissue available for culture, an alternative is to culture material from non-human primates as a closely related model. Whilst this has been more commonly used in other studies investigating follicle development, to the best of the authors' knowledge this has only very recently been deployed for the testing of chemotherapy-induced toxicity, in a study culturing slices of marmoset ovary.[14] Non-human primate ovary culture is a relatively new model, but could potentially be very useful in this regard, as it is the closest available species model to humans, with non-human primates having, for example, a similar protracted period of folliculogenesis *in vivo*.[9] Domestic species and ruminants are another alternative to the use of human tissue, but again, to the best of the authors' knowledge, they have not yet been used to test chemotherapy ovotoxicity.

For chemotherapy studies, culturing human ovarian tissue is clearly advantageous, eliminating concern about potential species differences that could exist in the tissue's response to chemotherapy treatment. However, very few studies have been conducted to investigate chemotherapy-induced toxicity *in vitro* using human tissue. As discussed above, the ideal model system will support primordial follicle growth, as this best reflects the ovarian reserve. The isolation of small follicles from human ovaries is difficult due to the density and toughness of the stromal tissue, and so culturing small slices of cortical tissue bypasses this issue and allows for the rapid activation of follicles. In cortical strips exposed to high concentrations of cyclophosphamide *in vitro*, primordial and early-stage follicles exhibited increased vacuolization and damage to granulosa cell nuclei and follicle basement membranes, as assessed by electron microscopy.[15] Doxorubicin increased apoptosis in primordial follicles and stromal cells along with damage to the microvasculature in cultured human cortical strips.[16] Specifically, doxorubicin induced DNA double-strand breaks in both human[16] and marmoset[14] ovary culture studies; the primate model was used further to investigate ovarian protection against chemotherapy-induced damage, showing the ability of dexrazoxane to help prevent this damage.[14]

6.5 RODENT MODEL STUDIES OF OVARIAN CHEMOTHERAPY TOXICITY

There are a range of different culture models established using rodent tissue spanning various stages of follicle development, and to date these have been the most frequently used model for studies to determine the toxicity of various chemotherapeutic agents *in vitro*. The culture of neonatal mouse ovaries is a popular technique for investigating chemotherapy-induced ovarian damage. These culture systems can span follicle formation and activation of primordial follicles through to the primary/secondary stage.[17] The small size of the tissue means that the ovary can be cultured whole, allowing for the maintenance of cell–cell and follicle interactions as well as integrity of the three-dimensional tissue structure. The vast majority of follicles within neonatal mouse ovaries are primordial. This makes neonatal mouse ovary culture a particularly useful model, allowing investigation specifically of the follicle stage of which the ovarian reserve is comprised. The relatively high numbers of follicles within the rodent ovary is also an advantage, providing a large amount of material for analysis.

A variety of different chemotherapeutic agents have been tested using neonatal ovary culture systems, including cisplatin, doxorubicin and cyclophosphamide. Cisplatin exposure caused extensive oocyte death in neonatal mouse ovaries as assessed by TAp63 accumulation, γH2AX expression and terminal deoxynucleotidyl transferase dUTP nick end labelling (TUNEL) staining,[18] and loss of primordial and early-growing follicles.[17,19] In contrast, doxorubicin exposure specifically caused granulosa cell damage in transitional and primary follicles, and primordial follicle loss.[17] Similarly, docetaxel specifically induced granulosa cell apoptosis in early-growing follicles, with damage to the oocyte a secondary effect.[20] Docetaxel also induced apoptosis in the stromal cell population of exposed ovaries in this model (Figure 6.2).[20] Cyclophosphamide metabolites caused a decrease in healthy primordial follicles and small primary follicles in mouse and rat ovaries, the most toxic of which was phosphoramide mustard, which specifically induced oocyte damage in primordial and early-growing follicles and granulosa cell damage in larger growing follicles.[21,22]

By examining effects of the chemotherapy drugs on primordial and early-growing follicles, the studies described above investigate the most important long-term consequence of chemotherapy exposure, namely

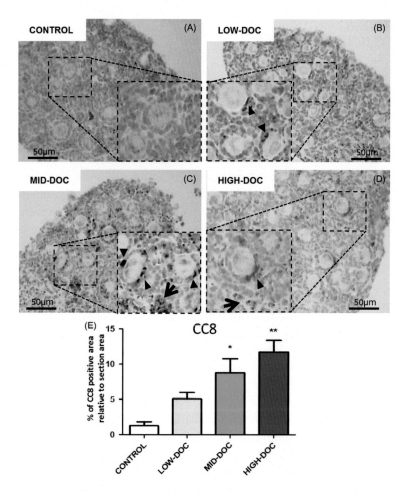

*Figure 6.2 **Analysis of an in vitro experiment to evaluate the effect of the chemotherapeutic agent docetaxal (DOC) on the ovary.** (A−D) Representative photomicrographs of mouse ovarian sections processed using immunohistochemistry to localize cleaved caspase 8 (CC8: brown), to illustrate the degree of apoptosis. (A) CONTROL ovary exhibiting little sign of CC8-positive apoptotic cells. (B−D) DOC-treated sections displaying marked apoptosis staining, with CC8 signal evident in many granulosa cells (GCs; arrowhead) and stromal cells (arrow). Sections were counterstained with haematoxylin (blue). Insets represent magnified images of respective framed areas. (E) Semi-automated analysis of immunohistochemistry. Exposure to DOC increased expression of CC8 relative to section area with the MID- and HIGH-DOC doses. Total of ovaries used in three independent experiments: CONTROL = 4, LOW-DOC = 4, MID-DOC = 6 and HIGH-DOC = 6. Kruskal−Wallis followed by Dunn post hoc tests were applied. Data are mean ± SEM. P < 0.05 (*), P < 0.01 (**) for DOC versus CONTROL. Taken from[20].*

the effect on the ovarian reserve. It is, though, also important to examine the effects of chemotherapy on later stages of folliculogenesis, as these will have the most immediate consequence to patients. Damage to follicles at later stages can also have an indirect impact on the ovarian reserve, due to the potential downstream impact on primordial follicles, as the loss of the production of inhibitory substances by growing follicles can lead to accelerated depletion of the primordial reserve, with more primordial follicles undergoing activation to replace damaged growing follicles. Individual follicle culture is a useful experimental tool that has been used extensively to study factors that impact upon late folliculogenesis. As follicles from the secondary/preantral stage can be grown in this system and oocytes fertilized to produce offspring,[23] this culture system is highly physiological. Limited studies have been conducted solely using individual follicle culture following chemotherapy exposure, although secondary follicles grown in the presence of the chemotherapeutic agent arsenic trioxide had a reduced growth rate compared to controls.[24]

Some *in vitro* research has been conducted exposing metaphase II (MII) oocytes to chemotherapy. As women undergoing chemotherapy are unlikely to conceive, this model is perhaps not as clinically relevant as other *in vitro* culture systems but may be of value in exploring the molecular pathways involved. Exposure of mature oocytes to doxorubicin caused DNA damage, activation of apoptotic pathways and associated morphological changes including budding and cellular fragmentation.[25,26] Etoposide treatment caused DNA double-strand breaks, as measured by γH2AX expression.[27] While this is of interest in determining the molecular mechanisms by which chemotherapy can cause cell death in oocytes, a limitation is that these oocytes are fully competent, and thus may react differently to insult than immature oocytes within follicles, as demonstrated by the increased vulnerability of germinal vesicle oocytes (obtained from antral follicles) to doxorubicin compared to that of MII oocytes.[28]

Given the highly physiological development of the mouse ovary *in vitro*, as demonstrated by the ability to generate live offspring, use of mouse *in vitro* studies allows for the possibility of offspring derived from *in vitro*-exposed oocytes to be used for transgenerational studies, including evaluation of epigenetic changes. Genetically modified tissue can also be used, meaning that different molecular mechanisms of cell death can be explored.[25] There are, however, several major issues which need to be borne in mind when using the rodent ovary as a model for humans. Rodents are polyovular and require only a short time for follicle development (3 weeks in the mouse compared to several months in humans). There is also very little stromal tissue in the mouse ovary in comparison to human ovaries. Rodent follicles could also have species-related differences in their response to chemo-therapeutic insults, as has been shown regarding the high sensitivity of mouse, compared to human, oocytes to radiation.[29] It is, therefore, important to check key findings from rodent studies using ovarian tissue from humans or large mammalian models.

6.6 OVARIAN CELL CULTURE TECHNIQUES IN CHEMOTHERAPY STUDIES

Both primary cell culture and culture of cell lines have been extensively used to evaluate the toxicity of different chemotherapeutic agents and the molecular mechanisms by which these agents cause cell death. Primary cell culture involves culturing cells harvested directly from an organ, and so these cells tend to have a limited lifespan (except for some tumour cells); in contrast, cell lines are immortalized, acquiring the ability to proliferate indefinitely. The majority of work has been conducted using cell lines derived from tumours, with occasional studies instead utilizing primary ovarian somatic cell culture. While these cell culture studies have many advantages, it is important to bear in mind that ovarian cells may have a different molecular response to treatment compared to cancer cells from other tissue types, and this could particularly be the case for female germ cells in meiotic arrest. Even where a cell line originates from ovarian tissue, its phenotype could have substantially altered during the cell line production process.

Work involving primary KK-15 granulosa cells indicates that doxo-rubicin exposure induces double-strand DNA breaks.[30] Primary rat granulosa cells exposed to paclitaxel undergo cell cycle arrest and subsequent apoptosis.[31] Both human and rat granulosa cells undergo cell death and reduced progesterone production following exposure to 4-hydroperoxycyclophosphamide, a metabolite of cyclophosphamide.[32] Whilst these results can be useful, such cells are lacking the specific cell interactions that exist within the three-dimensional follicle and tissue structure, and so may behave differently. Most drug doses that are used in these experiments are supraclinical; such pharmacological results may or may not reflect the physiological system.[33] In a single neuroblastoma cell line, for example, different dosages of doxorubicin can lead to concentration-dependant changes in activated cell death pathway including senescence, apoptosis and necrosis.[34]

6.7 CONCLUSION

The field of reproductive toxicology has shown that *in vitro* studies can make a substantial contribution to our understanding of the direct, detailed effects of toxicants on the ovary. Despite this, *in vitro* methods have, to date, been used relatively little in studies of how chemotherapy drugs damage the ovary, particularly regarding the use of human tissue or ovaries from other large mammalian models. This is an area ripe for future investigations.

REFERENCES

1. Stefansdottir A, Fowler PA, Powles-Glover N, Anderson RA, Spears N. Use of ovary culture techniques in reproductive toxicology. *Reprod Toxicol.* 2014;19(49C):117−135.

2. Pfitzer C, Chen CM, Wessel T, et al. Dynamics of fertility impairment in childhood brain tumour survivors. *J Cancer Res Clin Oncol.* 2014;140(10):1759−1767.

3. Colvin OM. An overview of cyclophosphamide development and clinical applications. *Curr Pharm Des.* 1999;5(8):555−560.

4. Madden JA, Hoyer PB, Devine PJ, Keating AF. Involvement of a volatile metabolite during phosphoramide mustard-induced ovotoxicity. *Toxicol Appl Pharmacol.* 2014;277(1):1−7.

5. Kawamura K, Cheng Y, Suzuki N, et al. Hippo signaling disruption and Akt stimulation of ovarian follicles for infertility treatment. *PNAS.* 2013;110(43):17474−17479.

6. Maiani E, Di Bartolomeo C, Klinger FG, et al. Reply to: Cisplatin-induced primordial follicle oocyte killing and loss of fertility are not prevented by imatinib. *Nat Med.* 2012;18:1172−1174.

7. Eppig JJ, O'Brien MJ. Development in vitro of mouse oocytes from primordial follicles. *Biol Reprod.* 1996;54(1):197−207.

8. O'Brien MJ, Pendola FL, Eppig JJ. A revised protocol for in vitro development of mouse oocytes from primordial follicles dramatically improves their developmental competence. *Biol Reprod.* 2003;68(5):1682−1686.

9. Telfer EE, Zelinski MB. Ovarian follicle culture: advances and challenges for human and nonhuman primates. *Fertil Steril.* 2013;99(6):1523−1533.

10. Telfer EE, McLaughlin M, Ding C, Thong KJ. A two-step serum-free culture system supports development of human oocytes from primordial follicles in the presence of activin. *Hum Reprod.* 2008;23(5):1151−1158.

11. Xu M, West-Farrell ER, Stouffer RL, Shea LD, Woodruff TK, Zelinski MB. Encapsulated three-dimensional culture supports development of nonhuman primate secondary follicles. *Biol Reprod.* 2009;81(3):587−594.

12. Comish PB, Drumond AL, Kinnell HL, et al. Fetal cyclophosphamide exposure induces testicular cancer and reduced spermatogenesis and ovarian follicle numbers in mice. *PLoS ONE.* 2014;9(4):e93311.

13. Chambers EL, Gosden RG, Yap C, Picton HM. In situ identification of follicle in ovarian cortex as a tool for quantifying follicle density, viability and developmental potential in strategies to preserve female fertility. *Hum Reprod.* 2010;25(10):2559−2568.

14. Salih SM, Ringelstetter AK, Elsarrag MZ, Abbott DH, Roti Roti EC. Dexrazoxane abrogates acute doxorubicin toxicity in marmoset ovary. *Biol Reprod.* 2015;92(3):1−11.

15. Raz A, Fisch B, Okon E, et al. Possible direct cytoxicity effects of cyclophosphamide on cultured human follicles: an electron microscopy study. *J Assist Reprod Genet.* 2002;19(10):500–506.

16. Soleimani R, Heytens E, Darzynkiewicz Z, Oktay K. Mechanisms of chemotherapy-induced human ovarian aging: double strand DNA breaks and microvascular compromise. *Aging.* 2011;3:8.

17. Morgan S, Lopes F, Gourley C, Anderson RA, Spears N. Cisplatin and doxorubicin induce distinct mechanisms of ovarian follicle loss; imatinib provides selective protection only against cisplatin. *PLoS ONE.* 2013;8(7):e70117.

18. Gonfloni S, Di Tella L, Caldarola S, et al. Inhibition of the c-Abl-TAp63 pathway protects mouse oocytes from chemotherapy induced death. *Nat Med.* 2009;15(10):1179–1185.

19. Kerr JB, Hutt KJ, Cook M, et al. Cisplatin-induced primordial follicle oocyte killing and loss of fertility are not prevented by imatinib. *Nat Med.* 2012;18:1170–1172.

20. Lopes F, Smith R, Anderson RA, Spears N. Docetaxel induces moderate ovarain toxicity in mice, primarily affecting granulosa cells of early growing follicles. *Mol Hum Reprod.* 2014;20 (10):948–959.

21. Desmeules P, Devine PJ. Characterizing the ovotoxicity of cyclophosphamide metabolites on cultured mouse ovaries. *Toxicol Sci.* 2006;90(2):500–509.

22. Petrillo SK, Desmeules P, Truong T-Q, Devine PJ. Detection of DNA damage in oocytes of small ovarian follicles following phosphoramide mustard exposures of cultured rodent ovaries *in vitro. Toxicol Appl Pharmacol.* 2011;253:94–102.

23. Spears N, Boland NI, Murray AA, Gosden RG. Mouse oocytes derived from in vitro grown primary ovarian follicles are fertile. *Hum Reprod.* 1994;9(3):527–532.

24. Ahn RW, Barrett SL, Raja MR, et al. Nano-encapsulation of arsenic trioxide enhances efficacy against murine lymphoma model while minimizing its impact on ovarian reserve in vitro and in vivo. *PLoS One.* 2013;8(3):e58491.

25. Perez GI, Knudson CM, Leykin L, Korsmeyer SJ, Tilly JL. Apoptosis-associated signalling pathways are required for chemotherapy-mediated female germ cell destruction. *Nat Med.* 1997;3(11):1228–1232.

26. Jurisicova A, Lee H-J, D'Estaing SG, Tilly J, Perez GI. Molecular requirements for doxorubicin-mediated death in murine oocytes. *Cell Death Differ.* 2006;13:1466–1474.

27. Marangos P, Carroll J. Oocytes progress beyond prophase in the presence of DNA damage. *Curr Biol.* 2012;22(11):989–994.

28. Bar-Joseph H, Ben-Aharon I, Rizel S, Stemmer SM, Tzabari M, Shalgi R. Doxorubicin-induced apoptosis in germinal vesicle (GV) oocytes. *Reprod Toxicol.* 2010;30(4):566–572.

29. Adriaens I, Smitz J, Jacquet P. The current knowledge on radiosensitivity of ovarian follicle development stages. *Hum Reprod Update.* 2009;15(3):359–377.

30. Roti Roti EC, Salih SM. Dextrazoxane ameliorates doxorubicin-induced injury in mouse ovarian cells. *Biol Reprod.* 2012;86(3):1–11.

31. Verga Falzacappa C, Timperi E, Bucci B, et al. T(3) preserves ovarian granulosa cells from chemotherapy-induced apoptosis. *J Endocrinol.* 2012;215(2):281–289.

32. Ataya KM, Pydyn EF, Ramahi-Ataya AJ. The effect of "activated" cyclophosphamide on human and rat ovarian granulosa cells in vitro. *Reprod Toxicol.* 1990;4(2):121–125.

33. Gewirtz DA. A critical evaluation of the mechanisms of action proposed for the antitumour effects of the anthracycline antibiotics adriamycin and daunorubicin. *Biochem Pharmacol.* 1999;57(7):727–741.

34. Rebbaa A, Zheng X, Chou PM, Mirkin BL. Caspase inhibition switches doxorubicin-induced apoptosis to senescence. *Oncogene.* 2003;22(18):2805–2811.

Strategies to Protect the Ovary

CHAPTER 7

Ovarian Tissue Cryopreservation for Fertility Preservation

Annette Klüver Jensen, Stine Gry Kristensen, and Claus Yding Andersen

Laboratory of Reproductive Biology, Copenhagen University Hospital — Rigshospitalet, University of Copenhagen, Copenhagen, Denmark

7.1 OVERVIEW

The survival rate among the ~2% of women of reproductive age who have suffered from invasive cancer has substantially increased.[1] Unfortunately, the necessary chemo- and radiotherapy carry the risk of unwanted side effects such as permanent infertility,[2] jeopardizing a woman's chances of having her own biological children.

Until recently, *in vitro* fertilization (IVF) and embryo transfer (ET) in combination with cryopreservation were considered the only options for women to conceive after recovery from a sterilizing cancer treatment. However, these methods carry some drawbacks and cannot sustain long-term fertility. They require controlled ovarian stimulation, and are not applicable to prepubertal girls. Cryopreservation and transplantation of ovarian tissue fulfils a number of these shortcomings.[3,4] Grafting of cryopreserved ovaria.;n tissue can restore menstrual cyclicity, does not require pretreatment and can be performed without delay. Further, the method does not require male gametes and is applicable even to prepubertal girls.

It is very difficult to give a patient an estimation of the risk of premature ovarian insufficiency (POI) due to the number of contributory factors, including age, disease, stage of disease and the fact that the planned chemotherapy treatment often changes during the course of treatment.[5] To help the physicians evaluate each patient, selection criteria such as the Edinburgh criteria (Figure 7.1)[6] can be used for guidance. However, these criteria should merely be used as guidance and not exact guidelines, as each woman should have an individual

Cancer Treatment and the Ovary. DOI: http://dx.doi.org/10.1016/B978-0-12-801591-9.00007-2

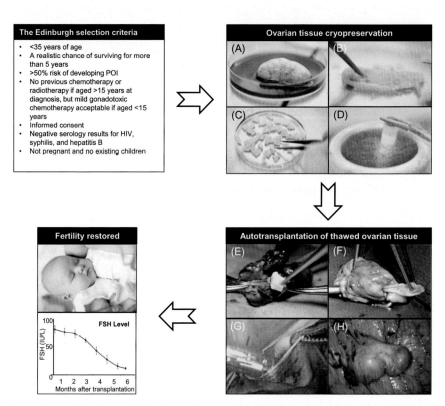

Figure 7.1 **Cryopreservation of ovarian tissue for fertility preservation.** *The Edinburgh selection criteria[6] can be used as a guideline for fertility counselling.* Cryopreservation of ovarian tissue: *(A) Oophorectomy; one ovary or part of an ovary is surgically removed. (B,C) Tissue processing; the medulla is removed and the cortex is trimmed to a thickness of 1–2 mm and cut into pieces of 5 × 5 mm. The pieces of cortex equilibrate in cold freezing solution for 30 minutes on ice. (D) Freezing; the cortex pieces are transferred to individual cryotubes and slow-frozen in liquid nitrogen.* Autotransplantation of thawed tissue: *(E) Orthotopic transplantation; thawed pieces of cortex are placed in subcortical pockets in the remaining ovary. (F) Cortical strips (5 × 15 mm) transplanted to the ovary. (G) Heterotopic transplantation; thawed pieces of cortex are placed in a subperitoneal pocket corresponding to the pelvic wall. (H) Two human antral follicles observed at a heterotopic graft site several years after transplantation.* Restored fertility: *Restoration of ovarian function, i.e., serum levels of follicle-stimulating hormone (FSH; in IU/L) in 12 Danish patients after autotransplantation of frozen-thawed ovarian tissue (mean ± SEM).* HIV, Human immunodeficiency virus; POI, premature ovarian insufficiency.

medical assessment of her ovarian reserve as well as her risk of POI. At the same time, she should have the opportunity to decide whether she is willing to take the risk of the operation in relation to the risk of becoming infertile. In Denmark, there is no exact upper age limit for women having ovarian tissue cryopreserved because some women have a better fertility potential than others (e.g., a high number of antral follicles) despite increased biological age (>35 years of age) and would therefore still benefit from having ovarian tissue cryopreserved. Furthermore, some women are still interested in having ovarian tissue stored even though their own risk of POI is fairly low, as they simply do not want to risk being unable to have children. Other women wish to have their ovary cryopreserved, even when their estimated chances of survival are vague, because the chance of survival is present. When the woman herself needs to cover the cost of cryopreservation, for example as this service is offered in USA, the situation becomes even more difficult. It is therefore of utmost importance to give these women information about their specific situation: their estimated chances of surviving the cancer and furthermore an estimation of their fertility potential, and to inform them that the current experience is rather limited and it is not known to what extent women of, for instance, advanced reproductive age may actually benefit from having tissue transplanted. A parallel situation is seen in IVF treatment, where women of advanced reproductive age receive treatment despite the fact that the success rate is rather low.

7.2 OVARIAN TISSUE CRYOPRESERVATION

The success of ovarian cryopreservation is based on the high cryopreservation tolerance of small primordial follicles compared to the larger growing follicles. The vast majority of primordial follicles are located in the outermost 1 or 2 mm of the ovarian cortex, which is relatively easy to isolate from the rest of the ovarian tissue (see Figure 7.1). The frozen ovarian cortex is stored in liquid nitrogen, allowing time for the patient to recover after cancer treatment is completed.

The most widely used protocol for ovarian cryopreservation is the slow-freezing method, and up until now all live births in humans, except for one birth from Japan,[7] have been achieved after slow-freezing.[8] The most used media compositions include either dimethyl sulfoxide (DMSO) or ethylene glycol, both in combination with sucrose.[4] However, vitrification of ovarian tissue could be one way of improving outcomes after freezing and re-implantation, as new results from non-human primates have shown superior follicular viability and development of antral follicles.[9-13] Despite these encouraging results, the slow-freezing method is still preferred over vitrification.

Currently there is no consensus as to how much of the ovarian cortex should be harvested for cryopreservation. It is recommended that oophorectomy should be performed on patients receiving high doses of alkylating agents, patients undergoing pelvic irradiation or total body irradiation, and younger prepubertal girls due to the small size of their ovaries.[8,14,15] For other patients, 4−5 ovarian cortical biopsies are harvested in some countries,[8,16] whereas a unilateral oophorectomy is carried out systematically in other countries.[17,18] The advantages of taking a whole ovary include minimizing the possibility of post-operational complications, e.g., bleeding, and also increasing the possibility of either grafting a large pool of follicles, which potentially should provide fertility with a higher efficacy, or performing repeated transplantation in case the tissue of the first transplantation becomes exhausted. However, the decision of excising one whole ovary in contrast to biopsies remains an issue of debate.

Transportation of ovarian tissue prior to cryopreservation allows hospitals without cryopreservation expertise to treat women locally for the cancer disease and just send the ovarian tissue to the centre that performs cryopreservation. This facilitates quality control, proper equipment, and personnel to fulfil clinical, legal and scientific standards required for proper conduction of the procedure. The feasibility of centralized cryobanking has been proven in Denmark and Germany, where ovarian tissue has been transported up to 20 hours prior to freezing,[4,19−21] and these principles are now being introduced in many other countries worldwide. The applicability of this approach will, however, be determined by local regulations.

7.3 TRANSPLANTATION OF CRYOPRESERVED OVARIAN TISSUE

Ovarian grafts transplanted to women who have entered menopause are able to re-establish a cyclic endocrine hormone milieu including preovulatory follicular development, appropriate conditions for conception, gestation and parturition, sometimes via IVF−ET.[3,4]

There are two main approaches for transplantation of human ovarian tissue (see Figure 7.1). In the orthotopic transplantation, a cortical fragment is grafted either to the remaining postmenopausal ovary, the broad ligament or to the ovarian fossa, whereas heterotopic transplantation refers to a graft site in a sub-peritoneal pocket outside the pelvic cavity, e.g., on the abdominal wall or subcutaneously, such as on a forearm. Most often, tissue is transplanted to the remaining *in situ* menopausal ovary either after decortication or by placing it under the cortex in order to facilitate a spontaneous pregnancy. Positioning of ovarian cortical pieces is most often performed with the original medulla-site inward and the original cortex-site outward in order to imitate the anatomically correct structure of the ovary. Grafting of the ovarian tissue to the anterior abdominal wall or to the lateral pelvic wall may be neces-sary if the remaining ovary is removed or has become very small. The general consensus is to transplant ovarian tissue to an orthotopic site, as it provides the optimal environment for follicular and corpus luteum development, compared to heterotopic sites: this means that temperature, pressure, paracrine factors and blood supply are similar to those observed during physiological conditions.[8] Even if transplanting ovarian tissue to heterotopic sites has some advantages,[22−23] only one live birth has been reported following this procedure.[24]

The size of the graft is also of importance in order to achieve a successful ovarian endocrine function. Meirow and co-workers illustrated two different surgical techniques, one with three ovarian tissue strips (5×15 mm) and a second technique with ovarian cubes in small cavities. Only the ovarian tissue strips regained endocrine function and led to a successful pregnancy with a live birth,[25] whereas the grafted small cubes did not become active. Since that study, a gen-eral consensus has emerged to re-implant pieces of 5×5 mm or even bigger tissue strips of 5×15 mm (see Figure 7.1E,F). According to our experience, larger strips ease the technical challenges of keeping the thawed tissue correctly in place at the graft site and allow firm, stable positioning of each individual piece of cortex, which is likely to facilitate revascularization.

Finally, the quantity of tissue transplanted depends on the desired function of the tissue. Over the years, experience and follow-up studies have developed a consensus to transplant less ovarian tissue for women who wish to regain their menstrual cycle with just one preovulatory follicle per cycle and more ovarian tissue for those who wish to become pregnant.

7.4 OUTCOME OF TRANSPLANTATIONS

On a worldwide basis it appears that at the time of writing 39 children have been born subsequent to transplantation (Table 7.1). Thirty of these cases have been described in peer-reviewed papers and 37 cases

Table 7.1 Number of Live Births Worldwide Resulting from Transplantation of Cryopreserved Ovarian Tissue

	Country	No. of Live Births	References
	Australia	3	24 +1*
	Belgium	7	30–34 +2*
	Denmark	8	18,26,35,36 +2*
	France	2	37,38
	Germany	4	20,21
	Israel	4	25,27,37,39
	Italy	1	40
	Japan	2	7 + 1*
	South Africa	1	1*
	Spain	3	41,42
	Sweden	1	1*
	USA	3	43,44

*Undocumented births

were mentioned in a review by Macklon and co-workers,[26] with additional children being reported since those publications.[27] Out of these live births, only one twin birth from Australia is a product of a heterotopic graft site.[24] A recent study from the authors' group described two additional pregnancies following heterotopic transplantation; however, one ended in early abortion and the other after premature preterm rupture of membranes at 19 weeks' gestation. The latter was probably due to the fact that the patient had received large doses of irradiation to the pelvis in connection with cancer treatment. It is emphasized that pregnancies from heterotopic grafted tissue require intensified luteal phase support and should be considered obstetrical high-risk pregnancies. In addition, two groups have been able to induce puberty by re-implanting frozen-thawed prepubertal ovarian tissue in two young girls.[28,29] This demonstrates the wide range of possibilities this method offers.

Three different European centres (from Belgium, Denmark and Spain) have collected and evaluated the results of 60 orthotopic re-implantation cases.[8] Fifty-one of the 60 patients had a follow-up >6 months after. The study showed that 93% of the women showed restoration of ovarian activity. Eleven of these 51 patients became pregnant and, at the time of the follow-up, six had already given birth to 12 healthy children.[8] In addition, >50% of the women who became pregnant were able to conceive naturally, which favours orthotopic transplantation. Moreover, the age of patients at the time of cryopreservation has previously been reported to be a predictive factor for pregnancy[37] and the majority of pregnant women were actually under the age of 30 years.

In three of the women who did not become pregnant no follicles were found in the grafted tissue, explaining why ovarian function did not resume. This highlights the importance of evaluating the presence of follicles before re-implantation. Ovarian activity is normally restored within 3.5 to 6.5 months post-grafting,[5] which concurs with the period of follicle growth from the primordial to the antral stage.[45]

In a recent study by Greve and co-workers, 12 women who had thawed ovarian tissue transplanted received assisted reproductive techniques (ART) and had 65 oocytes retrieved as a result of 72 cycles.[46] The outcome per cycle was a pregnancy rate of 6.9% and a live-birth rate of 3.8%. The low pregnancy rate was comparable with women 42 years old even though the oocytes originated from much younger women. It was suggested that the poor outcome reflected reduced follicular selection, rather than aged or damaged oocytes.[46] This explanation parallels the normal decline in female fecundity with increasing age.[46] Additionally, in this study, markers of follicle growth were in the majority of cases only found at very modest levels, as anti-Müllerian hormone remained undetectable in many patients, and levels of inhibin-B were found to be very low.[46,47] In a study by Dolmans and co-workers, four patients who were between the ages of 21–28 years at the time of cryopreservation had 21 oocytes retrieved.[48] The study found an empty follicle rate per retrieval of 29% (6/21), a tendency to a higher number of abnormal or immature oocytes and a lower embryo transfer rate, when compared to women who had not had their ovarian tissue cryopreserved.[48] Thus, the ovarian reserve is indeed very low following transplantation of thawed ovarian tissue and, at present, it is not possible to know whether the follicles have suffered from cryo injury or whether the low pregnancy rate reflects poor follicular selection.

In IVF treatment, the success rate is measured in the number of oocytes retrieved, the embryo transfer rate, the implantation rate and, of course, the clinical pregnancy rate per transferred cycle; however, these measurements are not always applicable to re-implanted cryopreserved ovarian tissue. Some women have ovarian tissue transplanted because they wish to avoid menopause and not because they have a wish to become fertile; some patients divorce their partner upon transplantation; and some women have a low, insufficient ovarian reserve in the remaining ovary. Furthermore, in contrast to IVF treatment, a pregnancy may occur as long as menstruation still occurs; the authors have, for instance, experienced pregnancies 5 years after grafting.[26]

7.5 LONGEVITY OF GRAFTS

In the study by Greve and co-workers, the longevity of functioning transplants was found to be between 9 months and 7 years.[46] Figure 7.1H shows two human antral follicles at a heterotopic graft site several years after transplantation of thawed ovarian tissue, demonstrating successful grafting and integration of the tissue. Other studies have shown that the transplanted grafts have a mean longevity of approximately 4−5 years when the follicular density was well preserved,[8,19,26,30] but follicular activity can persist for up to 7 years.[35] Given that the longevity of the tissue is good and that in many cases the women have enough tissue stored for 2−3 transplantations, the cryopreserved tissue could be enough to restore endocrine function until the natural age of menopause.

There are several important factors influencing the longevity of the grafted ovarian tissue: First of all, the follicular density of the transplanted tissue at the time of cryopreservation is of great importance. The follicular density depends, besides individual diversity, on the age of the patient at the time of cryopreservation, as the pool of follicles decreases with age.[49] Secondly, chemotherapy before cryopreservation of ovarian tissue has an impact on the ovarian reserve, depending on the gonadotoxicity of the chemotherapeutic agents.[2] Furthermore, the time between transplantation and revascularization of the tissue is critical for the number of surviving follicles. Baird and co-workers found that 60−70% of the follicles were lost in studies on transplantations in sheep.[50] Others have found that >50% of the primordial follicles are lost during and after ovarian transplantation.[51,52] This is mainly due to ischaemia in the tissue until angiogenesis has been established after transplantation.

Different publications have suggested that using angiogenetic and antiapoptotic factors (such as vascular endothelial growth factor or sphingosine-1-phosphate) prior to grafting could stimulate revascularization of the graft and thereby reduce ischaemia and oxidative stress.[8,53,54] Another important factor to facilitate angiogenesis is to fixate the grafted tissue properly to the graft site. This can be done by transplanting the thawed tissue in a subcortical or a subperitoneal pocket.[55] Microsurgical techniques can also be used to suture the graft onto the ovary after decortication followed by fixation of the graft with Interceed® or fibrin glue.[41,56] By keeping the grafted tissue in place and providing some pressure on the tissue, angiogenesis is believed to be enhanced, as it reduces the breakage of small blood vessels between the graft and graft site when the patient moves around.

7.6 RISK OF RE-IMPLANTING MALIGNANT CELLS

The most common diagnoses in girls, adolescents and young women undergoing cryopreservation of ovarian tissue include breast cancer, lymphoma, leukaemia and sarcoma. One serious concern when considering transplanting ovarian tissue is the risk of reintroducing malignant cells with the grafted tissue. However, cryopreservation of ovarian tissue is mainly offered to patients who have a good chance of surviving for >5 years, in which case most of these patients will have a low stage and limited disease and therefore a minimal risk of dissemination and ovarian involvement. Nevertheless, one exception is those patients who have haematological malignancies, where ovarian involvement cannot be excluded.

To minimize the risk of malignant cell contamination, various techniques can be used either separately or in combination: 1) During the laparoscopic oophorectomy the surgeon should look for possible gross pathology near and on the ovaries. 2) Before re-implantation, a piece of the frozen/thawed tissue can be evaluated by histology and immunohistochemistry (IHC) using the markers for the original tumour. 3) Evaluation by reverse transcription and quantitative polymerase chain reaction (RT-PCR and qPCR) can be carried out for specific cancer markers. 4) A piece of the frozen/thawed tissue can be transplanted under the skin of an immunodeficient mouse for a period of 4–6 months and evaluated for metastasis. However, even though one ovarian piece has been evaluated to be risk free, it is simply impossible to completely exclude that other pieces from the same patient would not be harbouring malignant cells and thereby lead to a relapse of the oncological disease.

Numerous studies have been undertaken on the risk of cell contamination by re-implanting cryopreserved ovarian tissue. In 2013, three independent groups carried out systematic reviews of the literature. All three reviews concluded that the highest risk of reintroducing cancer cells via autotransplantation of cryopreserved ovarian tissue was for leukaemic patients.[5,57–59] Additionally, some of the studies estimated that there was a moderate risk of cancer cell contamination for any of the gastrointestinal cancers.[5,58,59] The most reassuring data were found in relation to autotransplantation of patients surviving lymphoma. All three studies concluded that there was a low risk of metastasis in Hodgkin lymphoma,[5,57–59] although Dolmans and co-workers and Rosendahl and co-workers concluded that non-Hodgkin lymphoma could have a moderate risk of cell contamination and metastasis.[5,58,59]

Concerning patients with leukaemia, two *in vivo* studies from Denmark and Belgium have found that SCID-mice with xeno-transplanted tissue from patients in complete remission at the time of tissue collection did not develop leukaemia. These studies concluded that collection of ovarian tissue from patients with leukaemia should be done when they were in complete remission.[60,61] Even though these results are encouraging for continued storage of tissue from patients with a leukaemic diagnosis, more information is preferable prior to a large-scale transplantation activity in this group of patients.

Currently, it is unknown how few malignant cells can actually cause a relapse. We simply do not know what the minimal infection dose is. A few studies on rats have shown that even a few leukaemic cells could cause a relapse.[62,63] However, Lanman and co-workers carried out a study in 1950, which, although now considered ethically unacceptable, used American prison inmates, and showed that injections with huge amounts of malignant cells from one individual to another, did not cause malignant disease.[64] The results of this study may be applicable to leukaemic patients, who have often received bone marrow trans-plantation and thus new immune material from a different person.

Finally, a time perspective is also of importance, as many patients will not require grafting before maybe one or two decades. During this period, new techniques could evolve and be refined, e.g., *in vitro* maturation of primordial follicles or purging malignant cells from ovarian tissue and then transplantation of the isolated follicles. Therefore, young patients should not be denied the opportunity to have ovarian tissue cryopreserved due to uncertainties regarding the safety of the autotransplantation, but the parents should be informed of the current status of the technique and possible developments in the future. It is estimated that, to date, over 150 transplantations have been carried out worldwide. Among these transplantations, diagnoses such as leukaemia and sarcoma are included, and no relapses have been reported at any graft sites. However, one case report documents a relapse concerning the recurrence of a granulosa cell tumour, but no evidence of tumour was found at the graft site.[24] This could suggest that ovarian diseases may require stricter precautions.

7.7 CONCLUSION

Ovarian tissue cryopreservation is becoming a well-established technique for fertility preservation in multiple countries and 39 healthy children have so far been born worldwide through its application. However, the efficacy of this technique is difficult to evaluate, with some women interested in becoming pregnant but others wishing only to regain hormone production.

The number of transplantations is increasing as the women surviving their illness return to the clinic to get their fertility restored. Of the >150 cases worldwide (October 2014), ovarian activity is restored in the vast majority of cases. The longevity of the grafts is surprisingly good in some cases, lasting up to 7 years functioning effectively, thereby showing the strength of this technique.

Even though studies have indicated there may be a risk of re-implanting malignant cells to the patient in connection with the transplantation of thawed ovarian tissue, experience from the 150 + transplantation cases worldwide have not shown any relapses at the graft sites to date.

REFERENCES

1. Jemal A, Siegel R, Ward E, Murray T, Xu J, Thun MJ. Cancer statistics, 2007. *CA Cancer J Clin*. 2007;57:43−66.

2. Schmidt KT, Larsen EC, Andersen CY, Andersen AN. Risk of ovarian failure and fertility preserving methods in girls and adolescents with a malignant disease. *BJOG*. 2010;117:163−174.

3. Schmidt KT, Rosendahl M, Ernst E, et al. Autotransplantation of cryopreserved ovarian tissue in 12 women with chemotherapy-induced premature ovarian failure: the Danish experience. *Fertil Steril.* 2011;95:695−701.

4. Rosendahl M, Schmidt KT, Ernst E, et al. Cryopreservation of ovarian tissue for a decade in Denmark: a view of the technique. *Reprod Biomed Online.* 2011;22:162−171.

5. Donnez J, Dolmans MM. Fertility preservation in women. *Nat Rev Endocrinol.* 2013;9:735−749.

6. Wallace WHB, Smith AG, Kelsey TW, Edgar AE, Anderson RA. Fertility preservation for girls and young women with cancer: population-based validation of criteria for ovarian tissue cryopreservation. *Lancet Oncol.* 2014;15:1129−1136.

7. Kawamura K, Cheng Y, Suzuki N, et al. Hippo signaling disruption and Akt stimulation of ovarian follicles for infertility treatment. *Proc Natl Acad Sci USA.* 2013;110:17474−17479.

8. Donnez J, Dolmans MM, Pellicer A, et al. Restoration of ovarian activity and pregnancy after transplantation of cryopreserved ovarian tissue: a review of 60 cases of reimplantation. *Fertil Steril.* 2013;99:1503−1513.

9. Amorim CA, Curaba M, Van Langendonckt A, Dolmans MM, Donnez J. Vitrification as an alternative means of cryopreserving ovarian tissue. *Reprod Biomed Online.* 2011;23:160−186.

10. Amorim CA, Jacobs S, Devireddy RV, et al. Successful vitrification and autografting of baboon (Papio anubis) ovarian tissue. *Hum Reprod.* 2013;28:2146−2156.

11. Keros V, Xella S, Hultenby K, et al. Vitrification versus controlled-rate freezing in cryopreservation of human ovarian tissue. *Hum Reprod.* 2009;24:1670−1683.

12. Ting AY, Yeoman RR, Lawson MS, Zelinski MB. *In vitro* development of secondary follicles from cryopreserved rhesus macaque ovarian tissue after slow-rate freeze or vitrification. *Hum Reprod.* 2011;26:2461−2472.

13. Ting AY, Yeoman RR, Campos JR, et al. Morphological and functional preservation of pre-antral follicles after vitrification of macaque ovarian tissue in a closed system. *Hum Reprod.* 2013;28:1267−1279.

14. Jadoul P, Dolmans MM, Donnez J. Fertility preservation in girls during childhood: is it feasible, efficient and safe and to whom should it be proposed? *Hum Reprod Updat.* 2010;16:617−630.

15. Andersen CY, Kristensen SG, Greve T, Schmidt KT. Cryopreservation of ovarian tissue for fertility preservation in young female oncological patients. *Futur Oncol.* 2012;8:595−608.

16. Anderson RA, Wallace WH, Baird DT. Ovarian cryopreservation for fertility preservation: indications and outcomes. *Reproduction.* 2008;136:681−689.

17. Poirot CJ, Martelli H, Genestie C, et al. Feasibility of ovarian tissue cryopreservation for prepubertal females with cancer. *Pediatr Blood Cancer.* 2007;49:74−78.

18. Andersen CY, Rosendahl M, Byskov AG, et al. Two successful pregnancies following autotransplantation of frozen/thawed ovarian tissue. *Hum Reprod.* 2008;23:2266−2272.

19. Schmidt KL, Andersen CY, Loft A, Byskov AG, Ernst E, Andersen AN. Follow-up of ovarian function post-chemotherapy following ovarian cryopreservation and transplantation. *Hum Reprod.* 2005;20:3539−3546.

20. Dittrich R, Lotz L, Keck G, et al. Live birth after ovarian tissue autotransplantation following overnight transportation before cryopreservation. *Fertil Steril.* 2012;97:387−390.

21. Dittrich R., Nawroth F., Germeyer A., et al. P-295 39 transplantations of cryopreserved ovarian tissue − experience of the network FertiPROTEKT. ESHRE Conference 2014, Münich, Germany. 2014. p. i239.

22. Oktay K, Economos K, Kan M, Rucinski J, Veeck L, Rosenwaks Z. Endocrine function and oocyte retrieval after autologous transplantation of ovarian cortical strips to the forearm. *JAMA.* 2001;286:1490−1493.

23. Kim SS. Assessment of long term endocrine function after transplantation of frozen-thawed human ovarian tissue to the heterotopic site: 10 year longitudinal follow-up study. *J Assist Reprod Genet.* 2012;29:489–493.

24. Stern CJ, Gook D, Hale LG, et al. Delivery of twins following heterotopic grafting of frozen-thawed ovarian tissue. *Hum Reprod.* 2014;29:1828.

25. Meirow D, Levron J, Eldar-Geva T, et al. Pregnancy after transplantation of cryopreserved ovarian tissue in a patient with ovarian failure after chemotherapy. *N Engl J Med.* 2005;353:318–321.

26. Macklon KT, Jensen AK, Loft A, Ernst E, Andersen CY. Treatment history and outcome of 24 deliveries worldwide after autotransplantation of cryopreserved ovarian tissue, including two new Danish deliveries years after autotransplantation. *J Assist Reprod Genet.* 2014;31(11):1557–1564.

27. Shapira M, Raanani H, Cohen Y, Meirow D. Fertility preservation in young females with hematological malignancies. *Acta Haematol.* 2014;132:400–413.

28. Poirot C, Abirached F, Prades M, Coussieu C, Bernaudin F, Piver P. Induction of puberty by autograft of cryopreserved ovarian tissue. *Lancet.* 2012;379:588.

29. Ernst E, Kjaersgaard M, Birkebaek NH, Clausen N, Andersen CY. Case report: stimulation of puberty in a girl with chemo- and radiation therapy induced ovarian failure by transplantation of a small part of her frozen/thawed ovarian tissue. *Eur J Cancer.* 2013;49:911–914.

30. Donnez J, Squifflet J, Jadoul P, et al. Pregnancy and live birth after autotransplantation of frozen-thawed ovarian tissue in a patient with metastatic disease undergoing chemotherapy and hematopoietic stem cell transplantation. *Fertil Steril.* 2011;95(1787):e1–e4.

31. Donnez J, Dolmans MM, Demylle D, et al. Livebirth after orthotopic transplantation of cryopreserved ovarian tissue. *Lancet.* 2004;364:1405–1410.

32. Demeestere I, Simon P, Emiliani S, Delbaere A, Englert Y. Fertility preservation: successful transplantation of cryopreserved ovarian tissue in a young patient previously treated for Hodgkin's disease. *Oncologist.* 2007;12:1437–1442.

33. Demeestere I, Simon P, Moffa F, Delbaere A, Englert Y. Birth of a second healthy girl more than 3 years after cryopreserved ovarian graft. *Hum Reprod.* 2010;25:1590–1591.

34. Donnez J, Jadoul P, Pirard C, et al. Live birth after transplantation of frozen-thawed ovarian tissue after bilateral oophorectomy for benign disease. *Fertil Steril.* 2012;98:720–725.

35. Andersen CY, Silber SJ, Bergholdt SH, Jorgensen JS, Ernst E. Long-term duration of function of ovarian tissue transplants: case reports. *Reprod Biomed Online.* 2012;25:128–132.

36. Ernst E, Bergholdt S, Jorgensen JS, Andersen CY. The first woman to give birth to two children following transplantation of frozen/thawed ovarian tissue. *Hum Reprod.* 2010;25:1280–1281.

37. Donnez J, Silber S, Andersen CY, et al. Children born after autotransplantation of cryopreserved ovarian tissue. a review of 13 live births. *Ann Med.* 2011;43:437–450.

38. Roux C, Amiot C, Agnani G, Aubard Y, Rohrlich PS, Piver P. Live birth after ovarian tissue autograft in a patient with sickle cell disease treated by allogeneic bone marrow transplantation. *Fertil Steril.* 2010;93(2413):e15–e19.

39. Revel A, Laufer N, Ben Meir A, Lebovich M, Mitrani E. Microorgan ovarian transplantation enables pregnancy: a case report. *Hum Reprod.* 2011;1097–1103.

40. Revelli A, Marchino G, Dolfin E, et al. Live birth after orthotopic grafting of autologous cryopreserved ovarian tissue and spontaneous conception in Italy. *Fertil Steril.* 2013;99:227–230.

41. Sanchez-Serrano M, Crespo J, Mirabet V, et al. Twins born after transplantation of ovarian cortical tissue and oocyte vitrification. *Fertil Steril*. 2010;93(268):e11–e13.

42. Callejo J, Salvador C, Gonzalez-Nunez S, et al. Live birth in a woman without ovaries after autograft of frozen-thawed ovarian tissue combined with growth factors. *J Ovarian Res*. 2013;6:33.

43. Silber S, Kagawa N, Kuwayama M, Gosden R. Duration of fertility after fresh and frozen ovary transplantation. *Fertil Steril*. 2010;94:2191–2196.

44. Silber SJ, DeRosa M, Pineda J, et al. A series of monozygotic twins discordant for ovarian failure: ovary transplantation (cortical versus microvascular) and cryopreservation. *Hum Reprod*. 2008;:1531–1537.

45. Gougeon A. Regulation of ovarian follicular development in primates: facts and hypotheses. *Endocr Rev*. 1996;17:121–155.

46. Greve T, Schmidt KT, Kristensen SG, Ernst E, Andersen CY. Evaluation of the ovarian reserve in women transplanted with frozen and thawed ovarian cortical tissue. *Fertil Steril*. 2012;97:1394–8, e1.

47. Janse F, Donnez J, Anckaert E, de Jong FH, Fauser BCJM, Dolmans M-M. Limited value of ovarian function markers following orthotopic transplantation of ovarian tissue after gonadotoxic treatment. *J Clin Endocrinol Metab*. 2011;96:1136–1144.

48. Dolmans M-M, Donnez J, Camboni A, et al. IVF outcome in patients with orthotopically transplanted ovarian tissue. *Hum Reprod*. 2009;24:2778–2787.

49. Wallace WHB, Kelsey TW. Human ovarian reserve from conception to the menopause. *PLoS One*. 2010;5:e8772.

50. Baird DT, Webb R, Campbell BK, Harkness LM, Gosden RG. Long-term ovarian function in sheep after ovariectomy and transplantation of autografts stored at -196 C. *Endocrinology*. 1999;140:462–471.

51. Dolmans M-M, Martinez-Madrid B, Gadisseux E, et al. Short-term transplantation of isolated human ovarian follicles and cortical tissue into nude mice. *Reproduction*. 2007;134:253–262.

52. Van Eyck A-S, Jordan BF, Gallez B, Heilier J-F, Van Langendonckt A, Donnez J. Electron paramagnetic resonance as a tool to evaluate human ovarian tissue reoxygenation after xenografting. *Fertil Steril*. 2009;92:374–381.

53. Soleimani R, Heytens E, Oktay K. Enhancement of neoangiogenesis and follicle survival by sphingosine-1-phosphate in human ovarian tissue xenotransplants. *PLoS One*. 2011;6:e19475.

54. David A, Dolmans M-M, Van Langendonckt A, Donnez J, Amorim CA. Immunohistochemical localization of growth factors after cryopreservation and 3 weeks' xenotransplantation of human ovarian tissue. *Fertil Steril*. 2011;95:1241–1246.

55. Tryde Schmidt KL, Yding Andersen C, Starup J, Loft A, Byskov AG, Nyboe Andersen A. Orthotopic autotransplantation of cryopreserved ovarian tissue to a woman cured of cancer – follicular growth, steroid production and oocyte retrieval. *Reprod Biomed Online*. 2004;8:448–453.

56. Silber SJ, Lenahan KM, Levine DJ, et al. Ovarian transplantation between monozygotic twins discordant for premature ovarian failure. *N Engl J Med*. 2005;353:58–63.

57. Bastings L, Beerendonk CC, Westphal JR, et al. Autotransplantation of cryopreserved ovarian tissue in cancer survivors and the risk of reintroducing malignancy: a systematic review. *Hum Reprod Updat*. 2013;19:483–506.

58. Rosendahl M, Greve T, Andersen CY. The safety of transplanting cryopreserved ovarian tissue in cancer patients: a review of the literature. *J Assist Reprod Genet*. 2013;30:11–24.

59. Dolmans MM, Luyckx V, Donnez J, Andersen CY, Greve T. Risk of transferring malignant cells with transplanted frozen-thawed ovarian tissue. *Fertil Steril.* 2013;99:1514–1522.

60. Greve T, Clasen-Linde E, Andersen MT, et al. Cryopreserved ovarian cortex from patients with leukemia in complete remission contains no apparent viable malignant cells. *Blood.* 2012;120:4311–4316.

61. Dolmans M-M, Marinescu C, Saussoy P, Van Langendonckt A, Amorim C, Donnez J. Reimplantation of cryopreserved ovarian tissue from patients with acute lymphoblastic leukemia is potentially unsafe. *Blood.* 2010;116:2908–2914.

62. Hou M, Andersson M, Eksborg S, Söder O, Jahnukainen K. Xenotransplantation of testicular tissue into nude mice can be used for detecting leukemic cell contamination. *Hum Reprod.* 2007;22:1899–1906.

63. Jahnukainen K, Hou M, Petersen C, Setchell B, Söder O. Intratesticular transplantation of testicular cells from leukemic rats causes transmission of leukemia. *Cancer Res.* 2001; 61:706–710.

64. Lanman JT, Bierman HR, Byron RL. Transfusion of leukemic leukocytes in man; hematologic and physiologic changes. *Blood.* 1950;5:1099–1113.

CHAPTER 8

Current Clinical Approaches to Protecting the Ovary: GnRH Analogues

Hakan Cakmak and Mitchell P. Rosen
Division of Reproductive Endocrinology and Infertility, Department of Obstetrics, Gynecology, and
Reproductive Sciences, University of California, San Francisco, CA, USA

8.1 INTRODUCTION

Cancer is not uncommon and no longer considered as an incurable disease among reproductive-age women. Indeed, cancer is estimated to occur in approximately 2% of women under the age 40.[1,2] Over the past decades, there has been a remarkable improvement in the survival rates due to the marked progress in detecting cancer at earlier stages and the improvement in treatment modalities. With improvements in treatment outcomes, 83% of women younger than 45 years diagnosed with cancer in the USA survived between 2002 and 2012.[3] As a consequence of the increase in the number of patients surviving cancer, greater attention has been focused on the delayed effects of cancer treatments on the quality of future life of the survivor.[4,5]

The treatment for most of the cancer types in reproductive-age women involves either removal of the reproductive organs or cytotoxic treatment (chemotherapy and/or radiotherapy) that may partially or definitively affect reproductive function.[6] The irreversible gonadotoxic effects of chemotherapeutic agents on the ovary are well documented, especially for anthracyclines and alkylating agents (e.g., cyclophosphamide, busulfan and ifosfamide), which are the backbone of chemotherapy for breast cancer, lymphomas, leukaemia and sarcomas[7,8] (see also Chapter 4). Pelvic radiation therapy also causes follicular destruction and less than 2 Gy of radiation can deplete at least 50% of the follicles.[9] In addition, exposure to 5–10 Gy of pelvic radiation results in premature ovarian insufficiency (POI) in many women.[6] The risk of ovarian failure following chemotherapy is highly correlated with the woman's age and ovarian reserve at the time of treatment, type and dosage of drug administered,

Cancer Treatment and the Ovary. DOI: http://dx.doi.org/10.1016/B978-0-12-801591-9.00008-4

and the duration of drug exposure.[7,10] After chemotherapy, the long-term incidence of amenorrhoea is at least 25% at age 30 years and is 50% in women aged 35–40 years, whereas most women over 40 years of age become amenorrhoeic, and their chances of restoring ovarian function is dismal.[4,11] In addition, temporary amenorrhoea post-chemotherapy, but not duration of amenorrhoea, predicted a trend toward increased rates of infertility.[12]

Currently, the most widely used and effective strategies for fertility preservation in cancer patients are oocyte and embryo cryopreservation, which require the patient to undergo controlled ovarian stimulation in preparation for oocyte retrieval. Historically, this treatment option was often associated with significant delays in starting cancer treatment, which led to anxiety on the part of both the patient and medical team. However, with a random start ovarian stimulation, there are now minimal delays,[13] but it is potentially costly and invasive.[14]

Another option, which has resulted in live births but is still considered experimental according to the American Society for Reproductive Medicine, is the cryopreservation of ovarian tissue.[15] With this procedure, ovarian tissue is surgically removed and cryopreserved. The tissue can then be re-transplanted into the patient at a later time. Although pregnancies have been achieved, the efficiency of this method is controversial[16,17]: this is discussed further in Chapter 7.

Another option for fertility preservation that does not require cryopreservation of reproductive cells/tissues involves medical treatments that may protect the ovary. The most widely used agents are gonadotropin-releasing hormone (GnRH) agonists. Multiple clinical studies suggest a potential protective effect in patients receiving gonadotoxic chemotherapy, while others demonstrate no benefit. However, there are no data to support the use of GnRH agonists to protect the ovary from radiotherapy.[18,19] In this chapter, we review the physiology of GnRH and the role of GnRH agonist co-treatment with chemotherapy for the protection of the ovary.

8.2 DISTRIBUTION AND ROLES OF THE GnRH/GnRH RECEPTOR SYSTEM

8.2.1 GnRH

GnRH is a decapeptide synthesized by specific neurons located in the arcuate nucleus and in the preoptic area of the hypothalamus, and is released into the portal blood in a pulsatile fashion. GnRH binds with high affinity to specific G protein-coupled receptors on the surface of gonadotrope cells in the anterior pituitary, and induces biosynthesis and secretion of follicle-stimulating hormone (FSH) and luteinizing hormone (LH). These hormones act directly on the ovary, stimulating steroidogenesis and gametogenesis.

Twenty-three different isoforms of GnRH have been identified in various vertebrate species.[20] All of these peptides consist of 10 amino acids and have a similar structure, with at least 50% sequence identity[21]; they have been named according to the species from which they were initially isolated.[21,22] In mammals, the hypophysiotropic GnRH that stimulates the hypophysiotropic gonadotropin release was first isolated in pigs[23] and is designated as GnRH-I (mammalian GnRH),[24] while an early evolved and highly conserved new isoform that was discovered in the chicken is designated as GnRH-II (chicken GnRH).[25,26] In addition, in mammals, a third isoform, the salmon GnRH, named GnRH-III, has been reported.[27] In the human genome, only the *GnRH-I* and *GnRH-II* have been found. In humans, the expression of both *GnRH-I* and *-II* messenger RNA (mRNA) has been demonstrated in somatic and gonadal tissues such as the placenta, ovary, endometrium, trophoblast and the fallopian tubes[28,29] in addition to in the hypothalamus.

8.2.2 GnRH Receptor

The GnRH receptor (GnRHR) is a member of the G protein-coupled serpent-like membrane receptors, which consist of seven hydrophobic transmembrane chains, connected to each other with extracellular and intracellular loops. Transmembrane chains participate in receptor activation and the transmission of signals, and intracellular loops are involved in the interaction with G proteins and also other proteins participating in the intracellular signal transmission. Upon GnRH binding, the GnRHR undergoes a conformational change and stimulates G proteins, which in turn produces downstream activation of several signalling cascades, mainly inositol 1,4,5-triphosphate (IP3) and diacylglycerol (DAG), MAPK and adenylyl cyclase pathways.[24,30] Two homologous GnRHR have been identified: type I and type II. The absence of the carboxyl terminal tail results in slow internalization of type I GnRHR, and prevents rapid desensitization of the receptor.[31,32] In humans, the conventional type I GnRHR is mainly expressed in gonadotropes in the anterior pituitary. However, it is also expressed in numerous extrapituitary tissues including placenta, breast, ovary, uterus, prostate and the corresponding cancer cells.[33] The *type II GnRHR* has been cloned in marmoset as well as in non-human primates.[34,35] This receptor does have the characteristic carboxyl terminal tail, which allows its rapid desensitization.[36] However, a full-length *type II GnRHR* mRNA is absent in humans, as the open reading frame from the putative human *type II GnRHR* gene is disrupted by a frame shift, resulting in a premature stop codon.[37]

8.2.3 GnRH and GnRHR Expression in the Ovary

In human ovaries, *GnRHR* mRNA expression was initially demonstrated in granulosa-luteal cells aspirated from preovulatory follicles of women undergoing infertility treatment with *in vitro* fertilization (IVF).[38] Later, other researchers confirmed the presence of mRNA encoding for *GnRHR* in human granulosa-luteal cells and corpus luteum.[39,40] However, it is noteworthy that the levels of *GnRHR* mRNA in the ovary are almost 200-fold lower than in the pituitary.[41]

GnRH-I, GnRH-II and GnRHR protein expressions were not detected by immunostaining in the follicles from the primordial to the early antral stage.[42] However, in preovulatory follicles, both forms of GnRH and their common receptor were immunolocalized, predominantly to the granulosa cell layer, whereas the theca interna layer was weakly positive. In the corpus luteum, significant levels of GnRH-I, GnRH-II and GnRHR were observed in granulosa-luteal cells but not in theca-luteal cells. Both GnRH isoforms and the type I GnRHR were also immunolocalized to the ovarian surface epithelium.[42]

8.2.4 Direct Effect of GnRH in the Ovary

Consistent with its expression in the ovary, where its receptor is also detected, a direct effect of GnRH on ovarian cells has been demonstrated in animal models and humans. GnRH-I was found to inhibit DNA synthesis *in vitro* and to induce apoptosis in rat granulosa cells.[43,44] In humans, GnRH-I has been reported to exert an inhibitory action on ovarian steroidogenesis, decreasing progesterone production. However, others have reported either a stimulatory effect or an absence of any effect.[45] GnRH-II, similar to GnRH-I, inhibits progesterone secretion in human granulosa-luteal cells. GnRH-II also downregulates the receptors of FSH and LH.[46] Apart from its effects on ovarian steroidogenesis, GnRH is also implicated in downregulation of cell proliferation and induction of apoptosis. GnRH-I has been suggested as a luteolytic factor, increasing the number of apoptotic luteinized granulosa cells.[47] GnRH-I increased the number of apoptotic human granulosa cells obtained during oocyte retrieval for IVF.[47] Both GnRH-I and GnRH-II act as negative autocrine/paracrine regulators of cell proliferation in ovarian epithelial cells.[48] Therefore, in addition to its essential function of stimulating gonadotropin synthesis and secretion, GnRH seems to act as an autocrine and/or paracrine factor in the ovary, where it downregulates steroidogenesis and cell proliferation, and promotes apoptosis.

8.3 GnRH AGONISTS

GnRH agonists (e.g., goserelin, triptorelin, buserelin and leuprolide) are decapeptides, with similar structure to native GnRH and a great affinity to the GnRH receptors. After their binding to the receptors on gonado-tropes, they initially cause gonadotropin release (flare-up effect). After several days of continuous administration, this is followed by a dramatic drop in the circulating concentrations of FSH and LH, through a desensitization mechanism. GnRH agonists have greater affinity for the GnRHR than native GnRH; they also have greater resistance to enzymatic breakdown and a prolonged half-life compared to native GnRH (i.e., native GnRH has a half-life of about 2 to 4 minutes compared to 3 h for the GnRH agonist leuprolide).[49] The internalization of receptors and the slow dissociation of the agonist–receptor complex decrease the total number of functional GnRHR, leading to a prolonged desensitization.[50] During treatment with GnRH agonists, high concentrations of GnRH agonists circulate and bind all the available GnRHR, both in central and peripheral tissues.

However, studies on the effects of GnRH agonists on ovarian ste-roidogenesis have come to contradictory conclusions. Some have shown a lack of direct effect,[51–54] while others have reported either inhibition or stimulation of the production of oestrogen, progesterone, or both.[55–62] An interesting finding reported recently was that the treatment of mouse granulosa cells with GnRH agonists failed to increase in cAMP, phosphorylated ERK or phosphorylated ERK p38, which are downstream effectors of all G-coupled protein signalling cascades that GnRH is known to stimulate.[63] This controversial result can be explained by the presence of a different, yet unknown signalling pathway activated in the ovary by GnRH agonists.

8.4 GnRH AGONIST CO-TREATMENT WITH CHEMOTHERAPY FOR THE PROTECTION OF THE OVARY

8.4.1 Animal Studies

Based on the debated role of gonadal suppression in men in preserving testicular function against chemotherapy, and the belief that prepubertal girls are more resistant to gonadotoxic cancer treatment, the effect of GnRH agonists in preserving fertility by creating a prepubertal hormonal milieu has been investigated.

Animal studies have demonstrated a protective role of GnRH agonist treatment against chemotherapy-induced gonadal damage.[64,65] Primordial follicle loss associated with cyclophosphamide treatment was significantly lower in rhesus monkeys receiving GnRH agonist treatment compared with untreated animals (65% vs. 29%, respectively).[66] Interestingly, using the same model, a protective effect of GnRH agonist treatment was not observed after radiation therapy.[18] Oocytes are extremely sensitive to irradiation, and it seems unlikely that radiotherapy-induced gonadal damage can be prevented by gonadal suppression.[18,19]

In contrast to the protective effects of GnRH agonists, GnRH antagonists did not protect the ovary from chemotherapeutic damage. In a murine model, administration of GnRH antagonists not only did not prevent chemotherapeutic destruction of primordial follicles but also depleted primordial follicles, possibly through a direct effect on the ovary.[67] However, in another murine model, pretreatment with a GnRH antagonist significantly decreased the extent of primordial follicle damage induced by chemotherapy.[68] The discrepancy between these two studies may be due to the different treatment regimens or different strains of mice that were used.

8.4.2 Early Non-Randomized Studies in Human

Following encouraging findings in animal models, non-randomized studies with short term follow-up suggested a protective role for GnRH agonist co-treatment in women undergoing gonadotoxic chemotherapy.[69–76] These studies were criticized for their lack of randomization, lack of proper controls, possible recall bias, different follow-up periods for treatment and control groups, and not controlling for confounders. In addition, these studies most often had ovarian failure as an end-point, which may not reflect the decrease in primordial follicle count (partial injury), the most common impact in response to chemotherapy in young women.[77]

One of the largest studies, including 111 patients with Hodgkin lymphoma, was published by Blumenfeld et al.[78] In that study, a monthly depot injection of GnRH analogue was administered starting before chemotherapy for up to 6 months, in parallel to, and until the end of, chemotherapeutic treatment. The main significant difference was the rate of POI, which was 3% (2/65) in the GnRH agonist co-treatment group versus 37% (17/46) in the control group. In the GnRH agonist-treated group, 48 pregnancies occurred in 34 patients, who were 18−33 years old at chemotherapy administration, compared with 22 pregnancies in 16 patients in the control group who were 16−26 years old at chemotherapy administration.[78]

Both Huser et al. and Castelo-Branco et al. have reported similar results in patients with Hodgkin lymphoma in small case-control studies.[71,73] Significantly lower rates of POI were found in women using GnRH agonist with chemotherapy (10.0−20.8%) compared to those in control groups (71.1−76.9%).[71,73] These data suggest that the co-treatment of GnRH agonists in women undergoing chemotherapy for Hodgkin lymphoma may protect long-term ovarian function.

There are limited case-control studies evaluating the protective effect of GnRH agonists for other disease states. Recchia et al. investigated the protective effects on ovarian function and the efficacy and tolerability of goserelin, a GnRH analogue, added to adjuvant chemotherapy for early breast cancer.[79] After a median follow-up of 55 months, with 94% survival, 86% of the patients had resumed normal menses, and 84% were disease-free. The 1-, 3- and 5-year projected recurrence-free survival rates were 100%, 81% and 75%, respectively. Five years after treatment, one patient had a pregnancy that ended with a normal birth.[79] In another study, Recchia et al. retrospectively evaluated 100 consecutive premenopausal women with high-risk early breast carcinoma who received a GnRH agonist as ovarian protection during adjuvant chemotherapy.[80] After a median follow-up of 75 months, normal menses were resumed in all patients younger than age 40 years and by 56% of patients older than age 40 years. Three pregnancies were observed that resulted in two normal deliveries and one elective termination of pregnancy. These data show that the addition of GnRH agonists to adjuvant therapy of premenopausal patients with early breast cancer is well tolerated and may protect long-term ovarian function.

GnRH agonist co-treatment may also be beneficial for young women receiving cytotoxic chemotherapy for benign diseases. Somers *et al.* demonstrated that the treatment with GnRH agonist in parallel to cyclophosphamide therapy was associated with a significant reduction of POI in young women with severe systemic lupus erythematosus.[75] In their study, POI developed in one of 20 women treated with GnRH agonist (5%) compared with that in six of 20 controls (30%) matched by age and cumulative cyclophosphamide dose.[75]

8.4.3 Randomized Clinical Trials in Women

The protective effect seen in small observational studies of GnRH agonists and chemotherapy on future ovarian function triggered multiple larger and prospective randomized studies in recent years (Table 8.1). Most randomized trials to date have focused on whether GnRH agonists are protective in patients treated for breast cancer.

In a clinical trial undertaken by Badawy *et al.*,[81] 78 premenopausal breast cancer patients were randomly assigned to receive adjuvant chemotherapy (5-fluorouracil, doxorubicin (Adriamycin) and cyclophosphamide) with or without goserelin. The primary end-point of POI at 3−8 months post completion of chemotherapy demonstrated a significant benefit for patients receiving goserelin. Menstruation resumption was reported in 35 (90%) patients in goserelin/chemotherapy group versus 13 (33%) patients treated with chemotherapy alone ($p < 0.001$). However, in this study the rate of POI (67%) in the chemotherapy-alone group was higher than expected and the follow-up period was short. These may negatively affect the reliability and generalizability of the results. In addition, the pregnancy rate after treatment was not reported. Importantly, no information was given on either breast cancer hormonal status or tamoxifen use.

The ZIPP (Zoladex In Premenopausal Patients) randomized study was designed to compare the survival rate of different endocrine therapy regimens in premenopausal breast cancer patients treated with or without adjuvant chemotherapy and/or radiotherapy. A subset of

Table 8.1 Main Characteristics of the Randomized Controlled Trials

Trial	No. of Patients	Type of Cancer	Median/Mean Age (GnRH Agonist vs. Control)	Study Arms	Follow-up (Month)	Definition of POI	POI Incidence in GnRH-Treated Women (%)	POI Incidence in Controls (%)	Significant?
Badawy et al.[81]	78	Breast	30 vs. 29.2	CT + goserelin vs. CT	3 to 8	No resumption of spontaneous ovulation	11%	67%	Yes
Sverrisdottir et al.[82]	123	Breast	45 vs. 45	CT ± tamoxifen + goserelin vs. CT ± tamoxifen	36	Absence of menses	64% in gaserolin only, 93% in goserelin + tamoxifen	90% in CMF only, 87% in CMF + tamoxifen	Yes only for gaserolin only group
Del Mastro et al.[83]	281	Breast	39 vs. 39	CT + triptorelin vs. CT	12	No resumption of menstrual activity and postmenopausal levels of FSH and oestradiol	9%	26%	Yes
Song et al.[85]	183	Breast	40.3 vs. 42.1	CT + leuprolide vs. CT	12	No resumption of menses	40%	59%	Yes
Gerber et al.[87]	56	Breast	35 vs. 38.5	CT + goserelin vs. CT	24	No reappearance of two consecutive menstrual periods within 21–35 days	30%	43%	No
Munster et al.[88]	47	Breast	39 vs. 38	CT + triptorelin vs. CT	24	No resumption of menses	12%	10%	No

(Continued)

Table 8.1 (Continued)

Trial	No. of Patients	Type of Cancer	Median/ Mean Age (GnRH Agonist vs. Control)	Study Arms	Follow-up (Month)	Definition of POI	POI Incidence in GnRH-Treated Women (%)	POI Incidence in Controls (%)	Significant?
Elgindy et al.[89]	100	Breast	32 vs. 33	CT + triptorelin ± GnRH antagonist vs. CT alone	12	No resumption of menses	20% in "early CT" group, 16 % in "late CT" group	20% in both "early and late CT" groups	No
Moore et al.[90]	135	Breast	Not provided	CT + goserelin vs. CT	24	Amenorrhoea for the prior 6 months and post-menopausal FSH	8%	22%	Yes
Demeestere et al.[91]	84	Lymphoma	25.6 vs. 27.3	CT + oral contraceptives + triptorelin vs. CT + oral contraceptives	12	FSH level ≥ 40 IU/L	20%	19%	No

CMF, Cyclophosphamide, methotrexate and 5-fluorouracil; CT, chemotherapy; FSH, follicle-stimulating hormone.

patients was also enrolled in a prospectively planned sub-study assessing ovarian function.[82] Patients were randomized to one of four treatment arms: 1) control (no endocrine therapy); 2) goserelin; 3) tamoxifen; and 4) goserelin plus tamoxifen. All patients received endocrine therapy for 2 years, commenced concurrently with chemotherapy, regardless of oestrogen receptor status. In addition to endocrine therapy randomization, patients with node-positive disease received chemotherapy (cyclophosphamide, methotrexate and 5-fluorouracil), plus radiotherapy if four or more nodes were involved. From 260 assessable patients, 123 received six cycles of chemotherapy. At completion of endocrine therapy, amenorrhoea rates of 85% (control), 95% (tamoxifen), 97% (goserelin), and 92% (goserelin plus tamoxifen) were reported. Twelve months after completion of all therapy (36 months follow-up), there was a significant decrease in the amenorrhoea rate in patients treated with goserelin alone (64%) compared with persisting high rates for the other groups: 90% (control), 87% (tamoxifen), and 93% (goserelin + tamoxifen) ($p = 0.006$). Interestingly, no decrease in amenorrhoea was seen in the combined goserelin plus tamoxifen group. This contradictory finding was not explained in the study. In this study, the definition of premenopausal status (i.e., the last menstrual period within the preceding 6 months, including regular or irregular menstruation, without a specific age cut-off) allowed inclusion of perimenopausal patients. This likely contributed to the very high amenorrhoea rates reported across all groups. Pregnancy rates after treatment were not reported.

A phase III trial from Del Mastro et al.[83] assessed the efficacy of triptorelin in preventing chemotherapy-induced amenorrhoea when added to chemotherapy (cyclophosphamide, methotrexate and 5-fluorouracil, or anthracycline-based adjuvant or neoadjuvant chemotherapy) for breast cancer. With 281 enrolled patients, at a median follow-up of 12 months post-chemotherapy, the chemotherapy-induced amenorrhoea rate was significantly higher with chemotherapy alone (26%) compared with triptorelin/chemotherapy (9%) ($p < 0.001$). In this study, contradictory to the previously reported risk factors for POI,[84] only triptorelin treatment, but not age or chemotherapy type, was associated with the rate of chemotherapy-induced amenorrhoea rate. There were limited pregnancy outcomes following treatment. Twenty-four months after all patients had completed chemotherapy, one successful pregnancy in the control arm, and three pregnancies in the triptorelin group (resulting in two live births and one voluntary termination) were reported.

In a more recent phase II randomized trial, Song *et al.*[85] examined the efficacy of leuprolide acetate on preserving ovarian function in patients with breast cancer. A total of 183 patients were included in this prospective clinical trial and were assigned randomly to receive cyclophosphamide−doxorubicin-based chemotherapy only or chemotherapy plus leuprolide acetate. At the end of the 12 month follow-up, a significantly higher percentage of patients resumed menses in the chemotherapy plus leuprolide acetate group (53 out of 89) compared to patients in the chemotherapy-only group (39 out of 94) ($p < 0.05$).

At the time of writing, the OPTION (Ovarian Protection Trial In premenopausal breast cancer patients) randomized phase III trial has only been reported in abstract form.[86] This trial was designed to assess the ovarian protective effect of goserelin compared with no goserelin in patients with oestrogen receptor-negative early breast cancer treated with anthracyclinc- and/or cyclophosphamide-based adjuvant chemotherapy, with the primary end-point being amenorrhoea rate at 12 months after completion of chemotherapy. At the time of reporting, 173 patients had adequate follow-up data and 140 had adequate information about menstrual bleeding. No difference between treatment arms was reportedly found; however, data on actual amenorrhoea rates between arms have not yet been published.

The phase II ZORO (ZOledex Rescue of Ovarian function) trial included 56 premenopausal patients (28 patients in each arm) with oestrogen receptor-negative breast cancer randomly assigned to receive neoadjuvant anthracycline/cyclophosphamide-based chemotherapy with or without goserelin.[87] For the primary end-point of normal ovarian function, regular menstruation was reported in 70% versus 57% of patients treated with goserelin/chemotherapy versus chemotherapy alone, though this difference did not reach statistical significance. Similarly, there was no significant difference in the median time to resumption of menstruation: 6.1 versus 6.8 months for goserelin versus control group, respectively. There were limited pregnancy data available, with one pregnancy being reported in each treatment arm.

A randomized trial from Munster *et al.*[88] compared ovarian function in breast cancer patients treated with adjuvant anthracycline-based ± taxane chemotherapy with or without triptorelin. Ovarian function was defined as resumption of normal menstruation during follow-up of at least 2 years post-chemotherapy. Follow-up had been planned to extend to 5 years; however, the trial was terminated at interim analysis due to lack of effectiveness. Of the 49 patients enrolled, 27 patients received triptorelin/chemotherapy and 22 received chemotherapy alone. Most patients had oestrogen receptor-positive disease (73% and 74% for triptorelin and control groups, respectively). After a median follow-up of 18 months (range 5–43 months), resumption of menstruation was seen in 88% versus 90% for triptorelin versus control groups, respectively. Similarly, there was no significant difference in time to resumption of menstruation (5.8 vs. 5.0 months for triptorelin vs. control groups, respectively). Two pregnancies were reported, both in the control group. The number of patients attempting pregnancy was not given.

A phase II randomized trial from Elgindy et al.[89] was designed to assess the potential benefit of triptorelin in two different groups of premenopausal oestrogen receptor-negative early breast cancer patients. The "delayed chemotherapy" group (i.e., patients not requiring immediate cytotoxic treatment) were randomly assigned to chemotherapy plus triptorelin or chemotherapy alone. Patients requiring immediate commencement of chemotherapy were allocated to the "early chemotherapy" group, and randomly assigned to receive chemotherapy with or without triptorelin plus cetrorelix, a GnRH antagonist. The primary end-point of the rate of resumption of menstruation 12 months after chemotherapy was similar across the treatment groups (80% [triptorelin] vs. 84% [control] for the "delayed chemotherapy" group; and 80% [triptorelin + cetrorelix] vs. 80% [control] for the "early chemotherapy" group).

The POEMS trial[90] (Prevention of Early Menopause Study) included 214 women with hormone receptor-negative breast cancer under the age of 50 who were randomized to receive cyclophosphamide-containing treatment with or without monthly goserelin starting 1 week prior to the first chemotherapy dose. The primary study objective was to assess the rate of POI 2 years after chemotherapy, with POI defined as amenorrhoea during the previous 6 months along with postmenopausal concentrations of FSH. Only 135 patients had complete primary end-point data. The rate of POI was lower with goserelin (8%) compared to control group (22%) (OR,

0.3; 95% CI, 0.1−0.87; p = 0.03). The birth rate and number of pregnancies were significantly higher in goserelin group (18%) compared to control (9%) (OR, 2.22; 95% CI, 1.00−4.92; p = 0.05). During the follow-up, the patients treated with GnRH agonist also had improved disease-free and overall survival. The main strengths of the POEMS trial are the duration of follow-up with the demonstration of ovarian function resumption after 2 years and being the only study showing an increased pregnancy rate after using a GnRH agonist. The main limitation of the study is small sample size due to (a) early termination of the study due to lack of recruitment and financial support (the study had originally planned to randomize 400 patients), and (b) high dropout rate (38%) due to death or lack of FSH data.

There is only one randomized trial reported on whether GnRH agonists are protective in patients undergoing chemotherapy for lymphoma. In the study by Demeestere et al.[91] 84 patients were randomly assigned to receive either triptorelin plus norethisterone (GnRH agonist group) or norethisterone alone (control group) concomitantly with alkylating agent-based chemotherapy. After 1 year of follow-up, similar percentages of patients experienced POI in the GnRH agonist (20%) and control (19%) groups. More than half of patients in each group completely restored their ovarian function, but the anti-Müllerian hormone values were higher in the GnRH agonist group than in the control group (1.4 ± 0.35 vs. 0.5 ± 0.15 ng/mL, respectively; p = 0.04).

In summary, although animal and observational studies suggest that GnRH agonists may have protective effect in the ovary against chemotherapy, randomized trial outcomes are limited and inconsistent for breast cancer patients. Thus, the outcomes cannot be generalized to other cancer types. Despite a number of studies focused on one cancer type, these inconsistent results may stem from heterogeneity of study designs including cancer diagnosis, hormone receptor status in breast cancer, type and duration of the applied chemotherapeutic agents and GnRH agonists, the age of the patient population, definition of POI and follow-up period. The major issues in these clinical trials are small sample size, lack of long-term follow up and fertility/pregnancy information (except POEMS study). Also of concern is that, in most studies, there is lack of control for the confounding effects of tamoxifen, with tamoxifen being a known independent risk factor for amenorrhoea.[92] This tamoxifen effect was excluded in the ZORO, OPTION and POEMS studies by exclusive inclusion of patients with hormone receptor-negative tumours. It was hypothesized that in order for GnRH agonist to prevent ovarian damage, treatment should commence at least a week before chemotherapy. However, in several studies, some patients received GnRH agonist at the commencement of chemotherapy,[82,86] by which point potential efficacy might be diminished.

8.4.4 Meta-Analysis

Four recent meta-analyses have examined the efficacy of GnRH agonists on the resumption of menses after chemotherapy treatment for cancer, and again reported conflicting results due to inclusion of different studies in each meta-analysis. The first meta-analysis consisted of six trials that examined the reproductive outcomes among women with Hodgkin disease, ovarian cancer and breast cancer.[93] It suggested a potential benefit of GnRH agonist in premenopausal women undergoing chemotherapy for cancer, with significantly higher rates of spontaneous resumption of menses (OR, 3.46; 95% CI, 1.13–10.57) and ovulation (OR, 5.70; 95% CI, 2.29–14.20) in the GnRH agonist cohort compared with in the controls.[93] There was no statistically significant difference in the occurrence of spontaneous pregnancy (OR, 0.26; 95% CI 0.03–2.52). A recent meta-analysis consisting of nine trials that examined the efficacy of GnRH agonist, given before and during chemotherapy, for the prevention of POI in premenopausal women with Hodgkin disease, ovarian cancer and breast cancer also reported a significant reduction in the risk of POI (OR, 0.43; 95% CI, 0.22–0.84) in patients receiving GnRH agonist.[94] Another meta-analysis assessed the efficacy of GnRH agonist on the prevention of chemotherapy-related POI among breast cancer patients in the first year after treatment.[95] It included five randomized trials and found significantly fewer women treated with GnRH agonist to have POI compared with the placebo group (RR, 0.40; 95% CI, 0.21–0.75) but similar rates of menses resumption (RR, 1.31; 95% CI, 0.93–1.85) and spontaneous pregnancy (RR, 0.96; 95% CI, 0.20–4.56).[95] The most recent meta-analysis, evaluating the effect of concurrent use of GnRH agonists with chemotherapy on ovarian function in women with breast cancer who did not use tamoxifen, included four randomized trials.[96]

This meta-analysis failed to demonstrate any significant effect of GnRH agonists on the rate of resumption of spontaneous menses (OR, 1.47; 95% CI, 0.60–3.62).[96]

In these meta-analyses, the obvious lack of uniform chemotherapy regimens, follow-up duration, and POI definition negatively impacted accurate analyses. In addition, there was a major heterogeneity of the primary trials with regard to inclusion or exclusion of oestrogen receptor-positive breast cancer patients and treatment with tamoxifen. Treatment with tamoxifen is a significant potential source of bias in individual studies. The inconsistent age of patients included in the clinical trials may also have led to bias in these meta-analyses. Overall, even after these meta-analyses, the assessment of ovarian protection of GnRH agonists against chemotherapeutic agents is still inconclusive.

8.5 SUGGESTED MECHANISMS OF GONADOTOXIC PROTECTION BY GnRH AGONISTS

Alongside the clinical studies, there has been simultaneous investigation to determine if there is a biological basis for GnRH agonist treatment to provide a protective effect from the gonadotoxicity of chemotherapy.

One of the early hypotheses of how GnRH agonists might protect the ovary against chemotherapy-induced damage suggested that these agents could prevent accelerated follicle recruitment by inducing pituitary desensitization and thus preventing the increase in FSH concentrations.[70] Gonadotoxic chemotherapy may induce an accelerated rate of follicular recruitment and, in turn, follicular depletion, due to loss of growing follicles and thus low oestrogen/inhibin levels resulting in supraphysiological FSH.[70,97] However, primordial and primary follicles constitute the great majority of follicle reserve in adult ovary and do not express FSH receptors, and the early follicular recruitment is independent of FSH. Therefore, there is no molecular basis for this hypothesis. Reduced concentrations of local inhibitory factors (e.g., anti-Müllerian hormone) may contribute to increased recruitment of primordial follicles, although there is no direct evidence that this is relevant to any protective effect of GnRH agonists.

Another potential mechanism proposed to explain the beneficial effect of GnRH agonist on decreasing chemotherapy-associated gonadotoxicity is the decrease in utero-ovarian perfusion resulting from the hypo-oestrogenic state generated by pituitary–gonadal desensitiza-tion.[98,99] Decreased utero-ovarian perfusion may result in a lower total cumulative exposure of the ovaries to the chemotherapeutic agents, in turn resulting in a decreased gonadotoxic effect.[70]

It has been hypothesized that GnRH agonist may upregulate ovarian sphingosine-1-phosphate (S1P).[100] S1P is a pleiotropic lipid mediator of cell survival, and may be involved in prevention of chemotherapy-induced oocyte apoptosis.[101] S1P may prevent doxorubicin-induced oocyte death *in vitro*[102] and may suppress cyclophosphamide-induced apoptosis in human follicles.[103] However, this concept of GnRH agonist acting on S1P metabolism is speculative without direct evidence, and needs further evaluation.

The effects of GnRH analogues may also be explained through their direct actions on the ovary. GnRH receptor activation by their ligands may decrease cellular apoptosis.[104] In addition, Imai *et al.* have demonstrated *in vitro* that a GnRH analogue may decrease the gonadotoxic effect of chemotherapy (using doxyrubicin), independently of the hypogonadotropic milieu.[105] However, a recent study failed to demonstrate activation of downstream effectors of any G-coupled protein signalling cascade that GnRH is known to stimulate when mouse granulosa cells were treated with GnRH agonists,[63] leading to uncertainty about the functional activity of ovarian GnRHR.

Overall, current experimental data fail to explain how GnRH ago-nists could protect against gonadotoxic effects of chemotherapeutic agents. Until there is a biological mechanism identified, these molecu-lar findings may suggest that there is no protective effect of GnRH agonists: this could explain the inconsistent clinical data.

8.6 CONCLUSION

Fertility preservation in females diagnosed with cancer has become an important area of investigation due to increasing cancer survival rates combined with delayed childbearing. Initial studies using GnRH agonist co-treatment with chemotherapy have demonstrated promising results for fertility preservation. However, subsequent clinical trials and meta-analyses have reported conflicting results regarding the effectiveness of GnRH agonists. Based on current evidence, the role of GnRH agonists as ovarian protection agents remains controversial and unproven.

The heterogeneity across trials further impacts on interpretability of the data. Additional well-designed studies with larger populations and longer follow-up are needed before a definitive conclusion can be made about the protective effects of GnRH agonists against chemotherapy-induced ovarian damage.

Current evidence is limited by the fact that regular menstruation has generally been equated with fertility. However, resumption of menses may not be an accurate marker of fertility, since infertility and diminished ovarian reserve are observed in women who resume normal menstrual cycles after treatment with chemotherapy. The "gold standard" for assessing preserved fertility is successful pregnancy, although continuing ovarian activity through maintaining hormone production will also have health benefits. So far, limited data on pregnancy rate have been presented, primarily due to the prolonged follow-up required for this end-point.

Despite the controversy, difficult situations may arise where administration of GnRH agonists is reasonable. In these circumstances, given the absence of evidence of harm coupled with the potential negative long-term effects of infertility on the quality of life, consideration of GnRH agonists possibly protecting the ovary, but administered outside of a clinical trial setting may be warranted, provided there is careful discussion with the patient regarding the lack of proven benefit. GnRH agonist treatment can be offered as sole treatment or as an adjunct to other fertility preservation options including egg/embryo cryopreservation. One possible caveat is a theoretical risk of reducing the efficacy of chemotherapy when using GnRH agonists in a patient with oestrogen receptor-positive breast cancer, but there is no conclusive evidence for or against this hypothesis. As the data are unclear, GnRH agonists should only be used after a careful risk/benefit analysis.

REFERENCES

1. Georgescu ES, Goldberg JM, du Plessis SS, Agarwal A. Present and future fertility preservation strategies for female cancer patients. *Obstet Gynecol Surv.* 2008;63:725–732.

2. Jemal A, Bray F, Center MM, Ferlay J, Ward E, Forman D. Global cancer statistics. *CA Cancer J Clin.* 2011;61:69–90.

3. Howlader N, Noone AM, Krapcho M, et al., eds. SEER Cancer Statistics Review, 1975–2009 (Vintage 2009 Populations), National Cancer Institute. Bethesda, MD. Available at: <http://seer.cancer.gov/csr/1975_2009_pops09/>; Accessed 09.11.14.

4. Letourneau JM, Ebbel EE, Katz PP, et al. Pretreatment fertility counseling and fertility preservation improve quality of life in reproductive age women with cancer. *Cancer*. 2012;118:1710−1717.

5. Letourneau JM, Melisko ME, Cedars MI, Rosen MP. A changing perspective: improving access to fertility preservation. *Nat Rev Clin Oncol*. 2011;8:56−60.

6. Rodriguez-Wallberg KA, Oktay K. Options on fertility preservation in female cancer patients. *Cancer Treat Rev*. 2012;38:354−361.

7. Letourneau JM, Ebbel EE, Katz PP, et al. Acute ovarian failure underestimates age-specific reproductive impairment for young women undergoing chemotherapy for cancer. *Cancer*. 2012;118:1933−1939.

8. Maltaris T, Seufert R, Fischl F, et al. The effect of cancer treatment on female fertility and strategies for preserving fertility. *Eur J Obstet Gynecol Reprod Biol*. 2007;130:148−155.

9. Wallace WH, Thomson AB, Kelsey TW. The radiosensitivity of the human oocyte. *Hum Reprod*. 2003;18:117−121.

10. Morgan S, Anderson RA, Gourley C, Wallace WH, Spears N. How do chemotherapeutic agents damage the ovary? *Hum Reprod Update*. 2012;18:525−535.

11. Petrek JA, Naughton MJ, Case LD, et al. Incidence, time course, and determinants of menstrual bleeding after breast cancer treatment: a prospective study. *J Clin Oncol*. 2006;24:1045−1051.

12. Letourneau JM, Niemasik EE, McCulloch CE, et al. Temporary amenorrhea predicts future infertility in young women treated with chemotherapy. *J Cancer Ther Res*. 2013;2. Available from: http://dx.doi.org/10.7243/2049-7962-7242-7216.

13. Cakmak H, Katz A, Cedars MI, Rosen MP. Effective method for emergency fertility preservation: random-start controlled ovarian stimulation. *Fertil Steril*. 2013;100:1673−1680.

14. Cakmak H, Rosen MP. Ovarian stimulation in cancer patients. *Fertil Steril*. 2013;99:1476−1484.

15. Practice Committee of American Society for Reproductive M. Ovarian tissue cryopreservation: a committee opinion. *Fertil Steril*. 2014;101:1237−1243.

16. Donnez J, Dolmans MM, Demylle D, et al. Livebirth after orthotopic transplantation of cryopreserved ovarian tissue. *Lancet*. 2004;364:1405−1410.

17. Meirow D, Levron J, Eldar-Geva T, et al. Pregnancy after transplantation of cryopreserved ovarian tissue in a patient with ovarian failure after chemotherapy. *N Engl J Med*. 2005;353:318−321.

18. Ataya K, Pydyn E, Ramahi-Ataya A, Orton CG. Is radiation-induced ovarian failure in rhesus monkeys preventable by luteinizing hormone-releasing hormone agonists? Preliminary observations. *J Clin Endocrinol Metab*. 1995;80:790−795.

19. Jarrell JF, McMahon A, Barr RD, YoungLai EV. The agonist (d-leu-6,des-gly-10)-LHRH-ethylamide does not protect the fecundity of rats exposed to high dose unilateral ovarian irradiation. *Reprod Toxicol*. 1991;5:385−388.

20. Millar RP, Lu ZL, Pawson AJ, Flanagan CA, Morgan K, Maudsley SR. Gonadotropin-releasing hormone receptors. *Endocr Rev*. 2004;25:235−275.

21. Dubois EA, Zandbergen MA, Peute J, Goos HJ. Evolutionary development of three gonadotropin-releasing hormone (GnRH) systems in vertebrates. *Brain Res Bull*. 2002;57:413−418.

22. King JA, Millar RP. Evolutionary aspects of gonadotropin-releasing hormone and its receptor. *Cell Mol Neurobiol*. 1995;15:5−23.

23. Matsuo H, Baba Y, Nair RM, Arimura A, Schally AV. Structure of the porcine LH- and FSH-releasing hormone. I. The proposed amino acid sequence. *Biochem Biophys Res Commun*. 1971;43:1334−1339.

24. Sealfon SC, Weinstein H, Millar RP. Molecular mechanisms of ligand interaction with the gonadotropin-releasing hormone receptor. *Endocr Rev*. 1997;18:180–205.

25. Miyamoto K, Hasegawa Y, Nomura M, Igarashi M, Kangawa K, Matsuo H. Identification of the second gonadotropin-releasing hormone in chicken hypothalamus: evidence that gonadotropin secretion is probably controlled by two distinct gonadotropin-releasing hormones in avian species. *Proc Natl Acad Sci U S A*. 1984;81:3874–3878.

26. White SA, Bond CT, Francis RC, Kasten TL, Fernald RD, Adelman JP. A second gene for gonadotropin-releasing hormone: cDNA and expression pattern in the brain. *Proc Natl Acad Sci U S A*. 1994;91:1423–1427.

27. Montaner AD, Somoza GM, King JA, Bianchini JJ, Bolis CG, Affanni JM. Chromatographic and immunological identification of GnRH (gonadotropin-releasing hormone) variants. Occurrence of mammalian and a salmon-like GnRH in the forebrain of an eutherian mammal: *Hydrochaeris hydrochaeris* (Mammalia, Rodentia). *Regul Pept*. 1998;73:197–204.

28. Cheng CK, Leung PC. Molecular biology of gonadotropin-releasing hormone (GnRH)-I, GnRH-II, and their receptors in humans. *Endocr Rev*. 2005;26:283–306.

29. Millar RP. GnRHs and GnRH receptors. *Anim Reprod Sci*. 2005;88:5–28.

30. Cheung LW, Wong AS. Gonadotropin-releasing hormone: GnRH receptor signaling in extrapituitary tissues. *FEBS J*. 2008;275:5479–5495.

31. Blomenrohr M, Heding A, Sellar R, et al. Pivotal role for the cytoplasmic carboxyl-terminal tail of a nonmammalian gonadotropin-releasing hormone receptor in cell surface expression, ligand binding, and receptor phosphorylation and internalization. *Mol Pharmacol*. 1999;56:1229–1237.

32. Vrecl M, Heding A, Hanyaloglu A, Taylor PL, Eidne KA. Internalization kinetics of the gonadotropin-releasing hormone (GnRH) receptor. *Pflugers Arch*. 2000;439:R19–20.

33. Aguilar-Rojas A, Huerta-Reyes M. Human gonadotropin-releasing hormone receptor-activated cellular functions and signaling pathways in extra-pituitary tissues and cancer cells (Review). *Oncol Rep*. 2009;22:981–990.

34. Millar R, Lowe S, Conklin D, et al. A novel mammalian receptor for the evolutionarily conserved type II GnRH. *Proc Natl Acad Sci USA*. 2001;98:9636–9641.

35. Neill JD, Duck LW, Sellers JC, Musgrove LC. A gonadotropin-releasing hormone (GnRH) receptor specific for GnRH II in primates. *Biochem Biophys Res Commun*. 2001;282:1012–1018.

36. Millar RP. GnRH II and type II GnRH receptors. *Trends Endocrinol Metab*. 2003;14:35–43.

37. Morgan K, Conklin D, Pawson AJ, Sellar R, Ott TR, Millar RP. A transcriptionally active human type II gonadotropin-releasing hormone receptor gene homolog overlaps two genes in the antisense orientation on chromosome 1q.12. *Endocrinology*. 2003;144:423–436.

38. Peng C, Fan NC, Ligier M, Vaananen J, Leung PC. Expression and regulation of gonadotropin-releasing hormone (GnRH) and GnRH receptor messenger ribonucleic acids in human granulosa-luteal cells. *Endocrinology*. 1994;135:1740–1746.

39. Nathwani PS, Kang SK, Cheng KW, Choi KC, Leung PC. Regulation of gonadotropin-releasing hormone and its receptor gene expression by 17beta-estradiol in cultured human granulosa-luteal cells. *Endocrinology*. 2000;141:1754–1763.

40. Olofsson JI, Conti CC, Leung PC. Homologous and heterologous regulation of gonadotropin-releasing hormone receptor gene expression in preovulatory rat granulosa cells. *Endocrinology*. 1995;136:974–980.

41. Minaretzis D, Jakubowski M, Mortola JF, Pavlou SN. Gonadotropin-releasing hormone receptor gene expression in human ovary and granulosa-lutein cells. *J Clin Endocrinol Metab*. 1995;80:430–434.

42. Choi JH, Choi KC, Auersperg N, Leung PC. Differential regulation of two forms of gonadotropin-releasing hormone messenger ribonucleic acid by gonadotropins in human immortalized ovarian surface epithelium and ovarian cancer cells. *Endocr Relat Cancer.* 2006;13:641−651.

43. Billig H, Furuta I, Hsueh AJ. Gonadotropin-releasing hormone directly induces apoptotic cell death in the rat ovary: biochemical and in situ detection of deoxyribonucleic acid fragmentation in granulosa cells. *Endocrinology.* 1994;134:245−252.

44. Saragueta PE, Lanuza GM, Baranao JL. Inhibitory effect of gonadotrophin-releasing hormone (GnRH) on rat granulosa cell deoxyribonucleic acid synthesis. *Mol Reprod Dev.* 1997;47:170−174.

45. Olsson JH, Akesson I, Hillensjo T. Effects of a gonadotropin-releasing hormone agonist on progesterone formation in cultured human granulosa cells. *Acta Endocrinol (Copenh).* 1990;122:427−431.

46. Kang SK, Tai CJ, Nathwani PS, Leung PC. Differential regulation of two forms of gonadotropin-releasing hormone messenger ribonucleic acid in human granulosa-luteal cells. *Endocrinology.* 2001;142:182−192.

47. Zhao S, Saito H, Wang X, Saito T, Kaneko T, Hiroi M. Effects of gonadotropin-releasing hormone agonist on the incidence of apoptosis in porcine and human granulosa cells. *Gynecol Obstet Invest.* 2000;49:52−56.

48. Radovick S, Wondisford FE, Nakayama Y, Yamada M, Cutler Jr. GB, Weintraub BD. Isolation and characterization of the human gonadotropin-releasing hormone gene in the hypothalamus and placenta. *Mol Endocrinol.* 1990;4:476−480.

49. Chillik C, Acosta A. The role of LHRH agonists and antagonists. *Reprod Biomed Online.* 2001;2:120−128.

50. Ortmann O, Weiss JM, Diedrich K. Gonadotrophin-releasing hormone (GnRH) and GnRH agonists: mechanisms of action. *Reprod Biomed Online.* 2002;5(suppl 1):1−7.

51. Asimakopoulos B, Nikolettos N, Nehls B, Diedrich K, Al-Hasani S, Metzen E. Gonadotropin-releasing hormone antagonists do not influence the secretion of steroid hormones but affect the secretion of vascular endothelial growth factor from human granulosa luteinized cell cultures. *Fertil Steril.* 2006;86:636−641.

52. Casper RF, Erickson GF, Rebar RW, Yen SS. The effect of luteinizing hormone-releasing factor and its agonist on cultured human granulosa cells. *Fertil Steril.* 1982;37:406−409.

53. Lanzone A, Panetta V, Di Simone N, et al. Effect of gonadotrophin-releasing hormone and related analogue on human luteal cell function *in vitro. Hum Reprod.* 1989;4:906−909.

54. Weiss JM, Oltmanns K, Gurke EM, et al. Actions of gonadotropin-releasing hormone antagonists on steroidogenesis in human granulosa lutein cells. *Eur J Endocrinol.* 2001;144:677−685.

55. Bussenot I, Azoulay-Barjonet C, Parinaud J. Modulation of the steroidogenesis of cultured human granulosa-lutein cells by gonadotropin-releasing hormone analogs. *J Clin Endocrinol Metab.* 1993;76:1376−1379.

56. Dor J, Bider D, Shulman A, et al. Effects of gonadotrophin-releasing hormone agonists on human ovarian steroid secretion *in vivo* and *in vitro*-results of a prospective, randomized *in-vitro* fertilization study. *Hum Reprod.* 2000;15:1225−1230.

57. Gaetje R. Influence of gonadotrophin releasing hormone (GnRH) and a GnRH-agonist on granulosa cell steroidogenesis. *Clin Exp Obstet Gynecol.* 1994;21:164−169.

58. Guerrero HE, Stein P, Asch RH, de Fried EP, Tesone M. Effect of a gonadotropin-releasing hormone agonist on luteinizing hormone receptors and steroidogenesis in ovarian cells. *Fertil Steril.* 1993;59:803−808.

59. Miro F, Sampaio MC, Tarin JJ, Pellicer A. Steroidogenesis *in vitro* of human granulosa-luteal cells pretreated *in vivo* with two gonadotropin releasing hormone analogs employing different protocols. *Gynecol Endocrinol.* 1992;6:77−84.

60. Parinaud J, Beaur A, Bourreau E, Vieitez G, Pontonnier G. Effect of a luteinizing hormone-releasing hormone agonist (Buserelin) on steroidogenesis of cultured human preovulatory granulosa cells. *Fertil Steril.* 1988;50:597−602.

61. Pellicer A, Miro F. Steroidogenesis *in vitro* of human granulosa-luteal cells pretreated *in vivo* with gonadotropin-releasing hormone analogs. *Fertil Steril.* 1990;54:590−596.

62. Uemura T, Namiki T, Kimura A, Yanagisawa T, Minaguchi H. Direct effects of gonadotropin-releasing hormone on the ovary in rats and humans. *Horm Res.* 1994;41(suppl 1):7−13.

63. Torrealday S, Lalioti MD, Guzeloglu-Kayisli O, Seli E. Characterization of the gonadotropin releasing hormone receptor (GnRHR) expression and activity in the female mouse ovary. *Endocrinology.* 2013;154:3877−3887.

64. Ataya K, Moghissi K. Chemotherapy-induced premature ovarian failure: mechanisms and prevention. *Steroids.* 1989;54:607−626.

65. Bokser L, Szende B, Schally AV. Protective effects of D-Trp6-luteinising hormone-releasing hormone microcapsules against cyclophosphamide-induced gonadotoxicity in female rats. *Br J Cancer.* 1990;61:861−865.

66. Ataya K, Rao LV, Lawrence E, Kimmel R. Luteinizing hormone-releasing hormone agonist inhibits cyclophosphamide-induced ovarian follicular depletion in rhesus monkeys. *Biol Reprod.* 1995;52:365−372.

67. Danforth DR, Arbogast LK, Friedman CI. Acute depletion of murine primordial follicle reserve by gonadotropin-releasing hormone antagonists. *Fertil Steril.* 2005;83:1333−1338.

68. Meirow D, Assad G, Dor J, Rabinovici J. The GnRH antagonist cetrorelix reduces cyclophosphamide-induced ovarian follicular destruction in mice. *Hum Reprod.* 2004;19:1294−1299.

69. Blumenfeld Z, Shapiro D, Shteinberg M, Avivi I, Nahir M. Preservation of fertility and ovarian function and minimizing gonadotoxicity in young women with systemic lupus erythematosus treated by chemotherapy. *Lupus.* 2000;9:401−405.

70. Blumenfeld Z, von Wolff M. GnRH-analogues and oral contraceptives for fertility preservation in women during chemotherapy. *Hum Reprod Update.* 2008;14:543−552.

71. Castelo-Branco C, Nomdedeu B, Camus A, Mercadal S, Martinez de Osaba MJ, Balasch J. Use of gonadotropin-releasing hormone agonists in patients with Hodgkin's disease for preservation of ovarian function and reduction of gonadotoxicity related to chemotherapy. *Fertil Steril.* 2007;87:702−705.

72. Dann EJ, Epelbaum R, Avivi I, et al. Fertility and ovarian function are preserved in women treated with an intensified regimen of cyclophosphamide, adriamycin, vincristine and prednisone (Mega-CHOP) for non-Hodgkin lymphoma. *Hum Reprod.* 2005;20:2247−2249.

73. Huser M, Crha I, Ventruba P, et al. Prevention of ovarian function damage by a GnRH analogue during chemotherapy in Hodgkin lymphoma patients. *Hum Reprod.* 2008;23:863−868.

74. Pereyra Pacheco B, Mendez Ribas JM, Milone G, et al. Use of GnRH analogs for functional protection of the ovary and preservation of fertility during cancer treatment in adolescents: a preliminary report. *Gynecol Oncol.* 2001;81:391−397.

75. Somers EC, Marder W, Christman GM, Ognenovski V, McCune WJ. Use of a gonadotropin-releasing hormone analog for protection against premature ovarian failure during cyclophosphamide therapy in women with severe lupus. *Arthritis Rheum.* 2005;52:2761−2767.

76. Waxman JH, Ahmed R, Smith D, et al. Failure to preserve fertility in patients with Hodgkin's disease. *Cancer Chemother Pharmacol.* 1987;19:159–162.

77. Sonmezer M, Oktay K. Fertility preservation in female patients. *Hum Reprod Update.* 2004;10:251–266.

78. Blumenfeld Z, Avivi I, Eckman A, Epelbaum R, Rowe JM, Dann EJ. Gonadotropin-releasing hormone agonist decreases chemotherapy-induced gonadotoxicity and premature ovarian failure in young female patients with Hodgkin lymphoma. *Fertil Steril.* 2008;89:166–173.

79. Recchia F, Sica G, De Filippis S, Saggio G, Rosselli M, Rea S. Goserelin as ovarian protection in the adjuvant treatment of premenopausal breast cancer: a phase II pilot study. *Anticancer Drugs.* 2002;13:417–424.

80. Recchia F, Saggio G, Amiconi G, et al. Gonadotropin-releasing hormone analogues added to adjuvant chemotherapy protect ovarian function and improve clinical outcomes in young women with early breast carcinoma. *Cancer.* 2006;106:514–523.

81. Badawy A, Elnashar A, El-Ashry M, Shahat M. Gonadotropin-releasing hormone agonists for prevention of chemotherapy-induced ovarian damage: prospective randomized study. *Fertil Steril.* 2009;91:694–697.

82. Sverrisdottir A, Nystedt M, Johansson H, Fornander T. Adjuvant goserelin and ovarian preservation in chemotherapy treated patients with early breast cancer: results from a randomized trial. *Breast Cancer Res Treat.* 2009;117:561–567.

83. Del Mastro L, Boni L, Michelotti A, et al. Effect of the gonadotropin-releasing hormone analogue triptorelin on the occurrence of chemotherapy-induced early menopause in premenopausal women with breast cancer: a randomized trial. *JAMA.* 2011;306:269–276.

84. Valagussa P, Moliterni A, Zambetti M, Bonadonna G. Long-term sequelae from adjuvant chemotherapy. *Recent Results Cancer Res.* 1993;127:247–255.

85. Song G, Gao H, Yuan Z. Effect of leuprolide acetate on ovarian function after cyclophosphamide-doxorubicin-based chemotherapy in premenopausal patients with breast cancer: results from a phase II randomized trial. *Med Oncol.* 2013;30:667.

86. Leonard RC, Adamson D, Anderson R. The OPTION trial of adjuvant ovarian protection by goserelin in adjuvant chemotherapy for early breast cancer. *J Clin Oncol.* 2010;28:Abstr 590.

87. Gerber B, von Minckwitz G, Stehle H, et al. German Breast Group I. Effect of luteinizing hormone-releasing hormone agonist on ovarian function after modern adjuvant breast cancer chemotherapy: the GBG 37 ZORO study. *J Clin Oncol.* 2011;29:2334–2341.

88. Munster PN, Moore AP, Ismail-Khan R, et al. Randomized trial using gonadotropin-releasing hormone agonist triptorelin for the preservation of ovarian function during (neo) adjuvant chemotherapy for breast cancer. *J Clin Oncol.* 2012;30:533–538.

89. Elgindy EA, El-Haieg DO, Khorshid OM, et al. Gonadatrophin suppression to prevent chemotherapy-induced ovarian damage: a randomized controlled trial. *Obstet Gynecol.* 2013;121:78–86.

90 Moore HC, Unger JM, Phillips KA, Boyle F, Hitre E, Porter D, et al. Goserelin for ovarian protection during breast-cancer adjuvant chemotherapy. *N Engl J Med.* 2015;372 (10):923–932.

91. Demeestere I, Brice P, Peccatori FA, et al. Gonadotropin-releasing hormone agonist for the prevention of chemotherapy-induced ovarian failure in patients with lymphoma: 1-year follow-up of a prospective randomized trial. *J Clin Oncol.* 2013;31:903–909.

92. Walshe JM, Denduluri N, Swain SM. Amenorrhea in premenopausal women after adjuvant chemotherapy for breast cancer. *J Clin Oncol.* 2006;24:5769–5779.

93. Bedaiwy MA, Abou-Setta AM, Desai N, et al. Gonadotropin-releasing hormone analog cotreatment for preservation of ovarian function during gonadotoxic chemotherapy: a systematic review and meta-analysis. *Fertil Steril.* 2011;95:906−914:e901−904.

94. Del Mastro L, Ceppi M, Poggio F, et al. Gonadotropin-releasing hormone analogues for the prevention of chemotherapy-induced premature ovarian failure in cancer women: systematic review and meta-analysis of randomized trials. *Cancer Treat Rev.* 2014;40:675−683.

95. Yang B, Shi W, Yang J, et al. Concurrent treatment with gonadotropin-releasing hormone agonists for chemotherapy-induced ovarian damage in premenopausal women with breast cancer: a meta-analysis of randomized controlled trials. *Breast.* 2013;22:150−157.

96. Vitek WS, Shayne M, Hoeger K, Han Y, Messing S, Fung C. Gonadotropin-releasing hormone agonists for the preservation of ovarian function among women with breast cancer who did not use tamoxifen after chemotherapy: a systematic review and meta-analysis. *Fertil Steril.* 2014;102:808−815:e801.

97. Kalich-Philosoph L, Roness H, Carmely A, et al. Cyclophosphamide triggers follicle activation and "burnout"; AS101 prevents follicle loss and preserves fertility. *Sci Transl Med.* 2013;5:185ra162.

98. Kitajima Y, Endo T, Nagasawa K, et al. Hyperstimulation and a gonadotropin-releasing hormone agonist modulate ovarian vascular permeability by altering expression of the tight junction protein claudin-5. *Endocrinology.* 2006;147:694−699.

99. Saitta A, Altavilla D, Cucinotta D, et al. Randomized, double-blind, placebo-controlled study on effects of raloxifene and hormone replacement therapy on plasma no concentrations, endothelin-1 levels, and endothelium-dependent vasodilation in postmenopausal women. *Arterioscler Thromb Vasc Biol.* 2001;21:1512−1519.

100. Blumenfeld Z. How to preserve fertility in young women exposed to chemotherapy? The role of GnRH agonist cotreatment in addition to cryopreservation of embrya, oocytes, or ovaries. *Oncologist.* 2007;12:1044−1054.

101. Tilly JL. Commuting the death sentence: how oocytes strive to survive. *Nat Rev Mol Cell Biol.* 2001;2:838−848.

102. Morita Y, Perez GI, Paris F, et al. Oocyte apoptosis is suppressed by disruption of the acid sphingomyelinase gene or by sphingosine-1-phosphate therapy. *Nat Med.* 2000;6:1109−1114.

103. Meng Y, Xu Z, Wu F, et al. Sphingosine-1-phosphate suppresses cyclophosphamide induced follicle apoptosis in human fetal ovarian xenografts in nude mice. *Fertil Steril.* 2014;102:871−877:e873.

104. Grundker C, Emons G. Role of gonadotropin-releasing hormone (GnRH) in ovarian cancer. *Reprod Biol Endocrinol.* 2003;1:65.

105. Imai A, Sugiyama M, Furui T, Tamaya T, Ohno T. Direct protection by a gonadotropin-releasing hormone analog from doxorubicin-induced granulosa cell damage. *Gynecol Obstet Invest.* 2007;63:102−106.

CHAPTER 9

Preclinical Approaches to the Protection of Ovarian Function

Hadassa Roness and Dror Meirow

Center for Fertility Preservation, Sheba Medical Center, Tel Hashomer, Israel

9.1 INTRODUCTION

Recent advances in our understanding of the mechanisms underlying the impact of cytotoxic drugs on the ovary have opened up new directions for the protection of ovarian function from chemotherapy-induced damage. Studies are providing greater detail as to the pathways and factors triggered within the different cell types of the ovary by each drug class. As we gain increased knowledge of the specifics of these pathways, we reveal new targets for protective agents to reduce or prevent ovarian damage.

Most preclinical research on protective agents has concentrated on drug groups known clinically to have severe effects on follicle reserve, such as alkylating agents (cyclophosphamide, busulfan and dacarbazine), platinum complexes (cisplatin, carboplatin) and taxanes (paclitaxel). A large number of studies have also examined the anthracyclin antibiotic doxorubicin (DXR), which has been shown to affect the follicle reserve experimentally but which is currently considered of low clinical risk for premature ovarian insufficiency (POI).[1] Alkylating agents and platinum complexes work in a similar fashion, creating DNA crosslinks, which, in turn, cause DNA breaks, ultimately triggering apoptosis. The taxanes are microtubule-stabilizing agents, as distinct from DNA-damaging drugs, but it has been demonstrated that paclitaxel acts via Bax to induce apoptosis.[2,3] DXR is an intercalating agent that blocks DNA replication and causes double-stranded (ds) DNA breaks; it induces apoptosis primarily in the stroma and granulosa cells of growing follicles.[4,5] Within the oocyte, anthracyclin agents such as DXR have been shown to induce chromosomal fragmentation as well as fragmentation of the cytoplasm into apoptotic bodies.[6–8]

Cancer Treatment and the Ovary. DOI: http://dx.doi.org/10.1016/B978-0-12-801591-9.00009-6

9.2 OVARIAN PROTECTION BY AFFECTING APOPTOTIC PATHWAYS

The most extensively researched pathway in the context of chemotherapy- and radiotherapy-induced ovarian injury is the apoptotic pathway leading to cell death in response to DNA damage. Cytotoxic drug-induced apoptosis has been most clearly demonstrated in growing follicles (Figure 9.1B), and has been shown to originate in the proliferating granulosa cells.[9] Within the mature oocyte, apoptosis is mediated by several molecules including ceramide, Bax, and the caspases.[6–11] Ceramide has been identified as an initiator of DXR-induced apoptosis in oocytes,[11] and Bax, a protein produced by the Bcl2-gene family, plays an essential role in DXR-initiated apoptosis in mature mouse oocytes.[6,12] Caspases, members of the CASP protease family, have been universally shown to play a pivotal role as cell death effector molecules, and are central in the apoptotic pathway of mature oocytes.[12–14] Caspases-2, -12 and -3 specifically, have been shown to play a role in apoptosis of mature oocytes.[8,15] Mouse oocytes pretreated with a caspase inhibitor as well as mature oocytes derived from mice that lack expression of caspase-2 show a marked resistance to DXR-induced apoptosis.[6]

Investigation of apoptotic pathways within the dormant primordial follicle have revealed that p63, a homologue of anti-oncogene p53, found in the nucleus of oocytes, and specifically the p63 isoform TAp63, is a key mediator of the DNA damage apoptosis pathway in the response of primordial follicle oocytes to DNA injury.[16–18] The p63 pathway is upregulated when oocytes are exposed to external triggers of DNA damage such as radiation[18] and chemotherapy drugs such as cisplatin,[19] and loss of p63 in mouse oocytes results in resistance to the apoptotic effects of radiation[17] and cisplatin.[20] Tap63 activates apoptosis via proteins BAX (Bcl-2-associated X protein) and BAK (Bcl-2 homologous antagonist killer). The activation of apoptosis is mediated by TAp73,[20] either directly or through the activation of p53 upregulated modulator of apoptosis (PUMA) and phorbol-12-myristate-13-acetate-induced protein 1 (NOXA).[21] Oocyte-specific deletion of PUMA and/or both PUMA and NOXA in mice prevents γ-irradiation-induced apoptosis and can produce healthy offspring, indicating that the protected oocytes are capable of DNA repair and subsequent normal function.[21] Figure 9.1 schematically summarizes the key factors in the relevant apoptotic pathways.

(A)

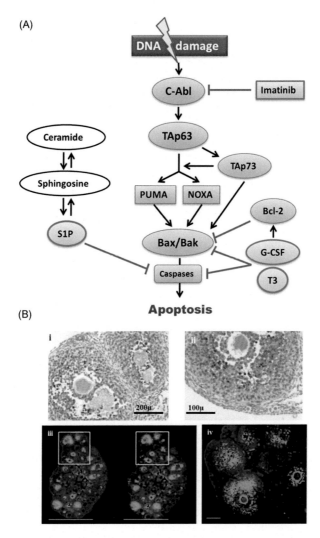

(B)

*Figure 9.1 (A) **Apoptosis pathways in the ovary and anti-apoptotic agents.** A schematic representation of the key molecules in the apoptotic pathway involved in chemotherapy-induced ovarian toxicity and the agents that act as apoptosis inhibitors (in purple). G-CSF, Granulocyte colony-stimulating factor; S1P, sphingosine-1-phosphate.*
*(B) **Apoptosis occurs in the granulosa cells of growing follicles.** Terminal deoxynucleotidyl transferase dUTP nick end labelling (TUNEL) staining showing apoptosis in growing follicles in mouse ovaries 24 hours after in vivo administration of 150 mg/kg cyclophosphamide (Cy) (i,ii), or 10 mg/kg doxorubicin (iii,iv). From[64] and[88].*

Each of these factors in the apoptotic cascade represents a potential target for blocking the effects of cytotoxic treatments on the ovary. One important concern with developing any agent that interferes with the apoptotic pathway as a potential protectant is that, since induction of apoptosis is a key anticancer action of many of the chemotherapy drug classes, interference with the apoptotic pathway may inhibit the anticancer effects of these drugs. A further concern is that blocking apoptosis in DNA-damaged oocytes could allow survival of germ cells that are beyond the repair capabilities of the DNA repair molecules. Fertilization of these genetically compromised oocytes could lead to an increased risk of miscarriage, fetal death or malformation.[22]

9.2.1 Imatinib

c-Abl protein tyrosine kinase has been shown to act as a "switch" for TAp63 transcriptional activity and the apoptotic pathway following exposure to the chemotherapy agents cisplatin[19] and DXR.[23] Other studies have further demonstrated that c-Abl plays a role in the maintenance of genomic integrity by dealing with DNA breaks in both meiotic and mitotic cells.[24,25] Imatinib is a competitive tyrosine kinase inhibitor clinically used in the treatment of cancer, most notably chronic myelogenous leukaemia. It was investigated as an agent to prevent primordial follicle loss caused by cisplatin[19] based on its role as a c-Abl kinase inhibitor. Results from that study demonstrated that co-administration of imatinib with cisplatin in mice reduced primordial follicle loss as well as improving fertility and reproductive outcomes. However, a subsequent study contested these results, finding that imatinib neither protected primordial follicle oocytes from cisplatin-induced apoptosis nor prevented loss of fertility in two independent strains of mice.[26] The conflicting results are likely due to a number of key differences in study design, but the role of c-Abl in the induction of oocyte degeneration has since been supported using another c-Abl inhibitor, GNF-2,[27] while others have confirmed the protective effect of imatinib.[20,28] An *in vitro* study[28] demonstrated that imatinib reduced the adverse effect of cisplatin on follicle health in cultured ovaries, while Kim et al.[20] showed that imatinib inhibited the cisplatin-induced upregulation of apoptotic mediators and reduced the damage to primordial follicles in a novel system of *in vitro* organ culture followed by subrenal grafting that enabled assessment of longer term effects.

Additional study is needed both to ascertain whether imatinib interferes with the anticancer action of cisplatin, since c-Abl has been shown to mediate cisplatin's action in a number of cancer cell lines,[29,30] and also to assess the genetic integrity of oocytes rescued from apoptosis.

9.2.2 Sphingosine-1-Phosphate

Sphingomyelin hydrolysis is a ceramide-promoted trigger of apoptotis and sphingosine-1-phosphate (S1P) is an inhibitor of this particular apoptotic pathway.[6,11] *In vivo* treatment of human ovarian tissue xenografts in mice with S1P increased vascular density and angiogenesis, and reduced follicle apoptosis.[31] *In vivo* administration of S1P before radiation exposure resulted in a dose-dependent preservation of follicle numbers in mice, and a virtually complete preservation of both primordial and growing follicles when S1P was administered at high doses.[11] In similar studies, S1P pretreatment was shown to reduce irradiation-induced primordial follicle depletion in rats,[32] primates,[33] and xenografted human ovarian tissue.[33]

S1P pretreatment demonstrated a protective effect in mice treated with dacarbazine,[34] reducing follicle loss and increasing pregnancy rates. Pretreatment and ongoing administration of S1P to mice carrying xenografted human ovarian tissue prevented the significant apoptosis of follicles caused by both cyclophosphamide (Cy) and DXR treatment.[35] *Ex vivo*, S1P treatment of mouse oocytes conferred resistance to DXR-induced apoptosis.[11] Results have not been uniformly positive, however, with one study reporting no reduction in Cy-induced follicle loss in S1P-treated rats.[32]

One limitation of S1P is that, because of its very short plasma half-life, it cannot be administered by systemic injection, and would either require continuous administration (studies used mini-osmotic pumps) or injections directly into the ovary. Local administration would, in theory, reduce the possibility that S1P would interfere with the therapeutic effects of chemotherapy drugs. Studies that have examined offspring derived from female mice and macaques that received S1P treatment prior to radiation showed no significant abnormalities of any kind,[33,36] possibly allaying concerns regarding persistent DNA damage in rescued oocytes, but no equivalent studies have been conducted after chemotherapy treatment.

9.2.3 Granulocyte Colony-Stimulating Factor

Granulocyte colony-stimulating factor (G-CSF) has been shown to significantly reduce the destruction of primordial follicles caused by Cy, busulfan and cisplatin,[37] as well as preventing damage to the microvessels and reducing markers of DNA damage in oocytes of early-growing follicles to control levels.[37] An extended mating study further demonstrated that mice that received G-CSF at the time of treatment produced significantly more pups per litter than those that received chemotherapy only. The mechanism of action of G-CSF requires clarification. It is possible that the protective effects of G-CSF on blood vessels decreases the chemotherapy-related blood vessel loss and the associated focal ischaemia shown by Meirow et al.[38] to be a contributing cause of follicle loss. However, the strongest hypothesis suggested by the authors is based on the anti-apoptotic effects of G-CSF. G-CSF has been shown to induce the expression of Bcl-2 and decrease Bax translocation, resulting in inhibition of apoptosis.[39,40]

While confirmation of these data and further examination of the mechanism behind the protective effect of G-CSF are still warranted, it claims certain advantages over some proposed attenuating agents as it is currently in clinical use in cancer patients for prevention of chemotherapy-induced neutropenia and has been shown not to reduce the efficacy of chemotherapeutic agents.

9.2.4 Thyroid Hormone (T_3)

Thyroid hormone, T_3, has been shown to have anti-apoptotic effects *in vitro* in a number of different cell lines,[41–43] including ovarian somatic and granulosa cells.[44–46] One study showed that T_3 reduced apoptosis in granulosa cells exposed *in vitro* to paclitaxel by downregulating caspase 3 and BAX[45]; however, another study showed no protective effect of T_3 when administered *in vitro* to mouse ovaries with docetaxol.[4]

9.2.5 Tamoxifen

Tamoxifen is an oestrogen receptor antagonist used in adjuvant treatment of hormone-sensitive cancers. Two rodent studies have proposed a protective effect when tamoxifen is given during chemo/radiotherapy. Ting et al. showed that *in vivo* treatment with tamoxifen significantly reduced follicle loss caused by Cy, as well as improving fertility outcomes post-treatment.[47] In addition, the study demonstrated that while *in vitro* DXR exposure increased oocyte fragmentation, the effect was reversed with tamoxifen. There are data to suggest that tamoxifen in conjunction with chemotherapy increases patient risk of long-term amenorrhoea; however, in the absence of chemotherapy, there was no association between use of tamoxifen and onset of long-term amenorrhoea.[48–54] Furthermore, since the effects of tamoxifen on the ovary are complicated and not well understood, amenorrhoea while on tamoxifen does not, in itself, indicate menopause.[55] The mechanism of a possible protective effect of tamoxifen against chemotherapy has not been investigated and is thought to relate either to its role as an oestrogen agonist with associated anti-apoptotic and anti-oxidant effects, or possibly its effects on the pituitary—gonadal axis.[56,57] Similarly, while whole body irradiation in rats significantly reduced primordial follicle numbers and anti-Müllerian hormone (AMH) serum levels, and increased oxidative stress markers, co-treatment with tamoxifen ameliorated all of these effects, as well as rescuing fertility post-treatment.[58] That study demonstrated that tamoxifen treatments resulted in a significant increase in both transcription and translation of insulin-like growth factor 1 (IGF-1), which has been shown to modulate gonadotropin action in the ovary by augmenting granulosa cell follicle-stimulating hormone (FSH) receptor expression and potentiating FSH action,[59] as well as to exert anti-oxidant and cryoprotective effects.[60,61]

It is worth noting here that it is difficult to explain how any effects on the pituitary—gonadal axis would protect the ovarian follicle reserve, since gonadotropins act only on the growing follicles, and primordial follicles are not directly under gonadotropin influence.[62,63] This is also discussed in Chapter 8 in relation to the potential protective role of GnRH agonist treatment.

9.3 PI3K FOLLICLE ACTIVATION PATHWAY AND OVARY PROTECTION

The most recent theory of chemotherapy-induced destruction of dormant follicles suggests that *in vivo*, chemotherapy agents such as Cy and cisplatin trigger an initiation of dormant follicle growth immediately following chemotherapy exposure, occurring simultaneously with large follicle apoptosis[64] (Figure 9.2). The activation of dormant follicles was shown to be mediated by an upregulation in the PI3K/PTEN/Akt signalling pathway, whose role in follicle quiescence has been well established by numerous knock-out mouse models[65] as well as its relevance to the human follicle by *in vitro* studies on cortical tissue.[66–68]

The route by which chemotherapy terminates follicle dormancy and induces activation of the PI3K/PTEN/Akt pathway may be via direct influence on the oocytes and pregranulosa cells of primordial follicles,

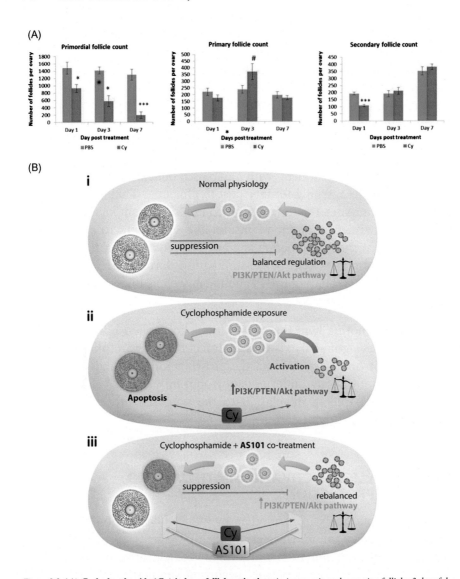

Figure 9.2 (A) Cyclophosphamide (Cy) induces follicle activation. An increase in early growing follicles 3 days following Cy treatment in 5-day old mice together with the increasing loss of dormant primordial follicles over 7 days demonstrates the phenomenon of follicle activation after Cy exposure. From[64]. PBS, phosphate-buffered saline.
(B) PI3k pathway role in follicle activation and burn out following chemotherapy. During normal follicular development, the ovary is in a state of equilibrium. The primordial follicles (PMFs) are under balanced regulation by the PI3K/PTEN/Akt pathway, and suppressive factors produced by growing follicles ensure that the vast majority of PMFs are maintained in a state of dormancy and very few are activated into growth. (ii) Cy upregulates the PI3K/PTEN/Akt follicle activation pathway, and causes apoptosis in growing follicles thereby reducing secretion of inhibitory factors. As a result, more PMFs are recruited into growth, develop and die, causing the reservoir to "burn out". (iii) AS101 prevents upregulation of the PI3K/PTEN/Akt follicle activation pathway, reduces apoptosis in the growing follicles, and restores the inhibitory influence on PMFs, returning the ovary to a state of equilibrium. The PMFs are maintained in a dormant state, fewer undergo recruitment and, as a result, the ovarian reserve is preserved. From[64].

or indirectly via chemotherapy-induced destruction of larger follicles.[69] Destruction of large follicles reduces negative regulation on the primordial follicle pool, thereby resulting in activation of primordial follicles in an attempt to replace the dying cohort of growing follicles. This has been suggested as the mechanism behind primordial follicle activation and depletion seen following exposure to carcinogen and ovotoxicant 3-methylcholanthrene.[70]

9.3.1 AS101

The immunomodulator AS101 is a non-toxic, tellurium-based compound that acts on the PI3K/PTEN/Akt pathway.[71,72] *In vivo* treatment of mice with AS101 reduced Cy-induced loss of primordial follicles as well as reducing apoptosis in granulosa cells of growing follicles and improving reproductive outcomes, via its modulation of the PI3K/PTEN/Akt pathway.[64] This mechanism could be by either direct protection or reduced activation of primordial follicles, or by a combination of both. No increase in fetal malformations was observed in mice whose fertility was protected by AS101 treatment, indicating that the functionality and genetic integrity of these rescued oocytes were not compromised. AS101 has been shown not to interfere with the primary anti-neoplastic activity of Cy *in vivo*, or with Cy metabolites *in vitro*.[73–76]

9.4 OTHER POTENTIAL METHODS OF REDUCING CHEMOTHERAPY-INDUCED OVOTOXICITY

9.4.1 Interference with Transport: Bortezomib

DXR has been shown to accumulate first in stromal cells, and progressively move towards the follicles,[5] so that DXR-induced dsDNA breaks occur first in stromal cells (within 2 hours of administration), followed by the granulosa cells (within 4 hours) that surround the oocytes.[5] In order to affect DNA damage, DXR must be transported across the nuclear membrane, via binding to proteasomes.[77] One study investigated the use of the high-affinity proteasome inhibitor bortezomib, which directly competes with DXR for binding to the proteasome, thus preventing DXR nuclear accumulation.[78] By decreasing DXR accumulation within ovaries, bortezomib pretreatment prevented DXR-induced DNA damage and the activation of downstream apoptotic pathways in all ovarian cell types, and resulted in improved reproductive outcomes in a rodent model. An inherent advantage of bortezomib is that it is already in clinical use as an anticancer agent.

9.4.2 Upregulation of Multidrug Resistance Gene (*MDR1*)

For drugs whose main mechanism of action is via the granulosa cells, another possibility for protection lies in the manipulation of phase III drug transporter enzymes such as MDR1. MDR1 is involved in the metabolism, elimination and detoxification of chemotherapy. Upregulation of *MDR1* has been linked to chemoresistance by promoting the transport and efflux of various lipid-soluble anticancer agents.[79,80] Upregulation of *MDR1* using retroviral transduction in a granulosa cell line was shown to protect cells from the toxic effects of DXR and paclitaxel in a dose-dependent manner.[81] *MDR1*-transduced granulosa cells showed significantly increased cell survival following treatment with either DXR or paclitaxel. Conversely, in a different study, inhibition of MDR transporters in mouse and human oocytes led to increased susceptibility to cyclophosphamide toxicity.[82]

A particular concern with this method of chemoprotection is that it utilizes retrovirus therapy, raising the possibility of viral gene transfer to germ cells and ultimately to potential fetuses.

9.4.3 Drug Encapsulation

Another direction that has been investigated to reduce the toxic side effects of chemotherapy, particularly for solid tumours, is drug encapsulation with/without tumour targeting. One successful example of this method is a pegylated (polyethylene glycol-coated) liposome-encapsulated form of DXR, which reduces the cardiotoxicity of the drug and is currently in clinical use for the treatment of ovarian cancer and multiple myeloma.[83] A recent study examined the possibility that nano-encapsulation of arsenic trioxide (As_2O_3; used in the treatment of haematological malignancies) may increase its efficacy against solid tumours while simultaneously reducing its effects on the ovary.[84] The study demonstrated that nano-encapsulation of As_2O_3 resulted in lower peak plasma levels, limited its tissue distribution and reduced its impact on ovarian and follicular function. While the study did not examine the impact of the encapsulated drug on ovarian reserve or function following treatment, the concept holds great potential and warrants additional investigation.

Table 9.1 Summary of Agents for the Prevention of Ovarian Chemotoxicity				
Protective Agent/ Method	Mechanism of Action on Ovary	Studies Demonstrating Protective Effect *In Vivo*	Studies Demonstrating Protective Effect *In Vitro*	Studies Demonstrating No Effect
Imatinib	Anti-apoptotic via inhibition of c-Abl kinase	Rodent: Gonfloni, 2009[19]	Rodent: Morgan, 2013[28]; Kim, 2013[20]	Rodent: Kerr, 2012[26]
S1P	Anti-apoptotic agent	Rodent: Jurisicova, 2006[85]; Hancke, 2007[34]; (Morita, 2000[11]*; Kaya, 2008[32]*)	Rodent: Morita, 2000[11]	Rodent: Kaya, 2008[32]
		Primate: (Zelinski, 2011[33]*)		
		Human xenograft: Li, 2014; (Zelinski, 2011[33]*)		
G-CSF	Unclear: reduces apoptosis via effects on Bcl2 and BAX, or neovascularization	Rodent: Skaznik-Wikiel, 2013[37]; Akdemir, 2014[86]		
Thyroid hormone T_3	Anti-apoptotic action	Rodent: Verga Falzacappa, 2012[45]		Rodent: Lopes, 2014[4]
Tamoxifen	Anti-apoptotic and anti-oxidant, possibly gonadal suppression	Rodent: Ting, 2010[47]; (Mahran, 2013[58]*)	Rodent: Ting, 2010[47]	Human: Sverrisdottir, 2009[87]
AS101	Modulation of PI3K/ PTEN/Akt follicle activation pathway	Rodent: Kalich-Philosoph, 2013[64]		
Bortezomib	Interferes with nuclear transport of DXR, prevents DNA damage	Rodent: Roti Roti, 2014[78]		
MDR1	Upregulation of MDR1 promoting efflux of lipid-soluble cancer drugs		Rodent: Salih, 2011[81]	
Drug encapsulation	Nano-encapsulation to limit tissue distribution	Rodent: Ahn, 2013[84]	Rodent: Ahn, 2013[84]	

DXR, Doxorubicin; G-CSF, granulocyte colony-stimulating factor; S1P, sphingosine-1-phosphate.
˙These studies examined radiation not chemotherapy.

9.5 CONCLUSION

Current options for female fertility preservation in the face of cytotoxic treatments include embryo, oocyte and ovarian tissue cryopreservation; however, these methods are limited by patient age, status and/or available timeframe before treatment, and also necessitate invasive procedures. Pharmacological agents that can prevent the loss of follicles at the time of treatment would provide significant advantages over existing fertility-preservation techniques in that they would be suitable for patients of all ages and life stages, would not require invasive surgical procedures or subsequent use of assisted reproductive technologies, and would prevent the myriad endocrine-related effects of POI other than infertility. Different cytotoxic agents act in specific ways on the different cell populations within the ovary, providing numerous targets for potential attenuating agents. Interference with any one of these pathways of injury has the potential to reduce the impact of cytotoxic agents on the ovary. Many of the agents discussed above act on the apoptotic pathway to prevent cell death (Table 9.1); however, it is also possible that prevention of apoptosis ties in with the other pathway discussed – the follicle-activation pathway. Apoptosis and death of the growing follicles removes the negative regulation normally provided by AMH secretion from these follicles. Thus, any agent that prevents apoptosis in growing follicles will also indirectly prevent dormant follicle activation and loss.

The protective agents discussed here are in very preliminary stages of study, but while there may be a long path from lab to clinic, these initial studies continue to advance our understanding of the mechanisms and pathways involved in cytotoxic ovarian damage, thus bringing us closer to preventing chemotherapy-induced infertility.

REFERENCES

1. Meirow D, Biederman H, Anderson RA, Wallace WH. Toxicity of chemotherapy and radiation on female reproduction. *Clin Obstet Gynecol.* 2010;53:727–739.

2. Fan W. Possible mechanisms of paclitaxel-induced apoptosis. *Biochem Pharmacol.* 1999;57:1215–1221.

3. Stumm S, Meyer A, Lindner M, Bastert G, Wallwiener D, Guckel B. Paclitaxel treatment of breast cancer cell lines modulates Fas/Fas ligand expression and induces apoptosis which can be inhibited through the CD40 receptor. *Oncology.* 2004;66:101–111.

4. Lopes F, Smith R, Anderson RA, Spears N. Docetaxel induces moderate ovarian toxicity in mice, primarily affecting granulosa cells of early growing follicles. *Mol Hum Reprod.* 2014;20:948–959.

5. Roti Roti EC, Leisman SK, Abbott DH, Salih SM. Acute doxorubicin insult in the mouse ovary is cell- and follicle-type dependent. *PLoS One.* 2012;7:e42293.

6. Perez GI, Knudson CM, Leykin L, Korsmeyer SJ, Tilly JL. Apoptosis-associated signaling pathways are required for chemotherapy-mediated female germ cell destruction. *Nat Med.* 1997;3:1228–1232.

7. Perez GI, Tao XJ, Tilly JL. Fragmentation and death (a.k.a. apoptosis) of ovulated oocytes. *Mol Hum Reprod.* 1999;5:414–420.

8. Bar-Joseph H, Ben-Aharon I, Rizel S, Stemmer SM, Tzabari M, Shalgi R. Doxorubicin-induced apoptosis in germinal vesicle (GV) oocytes. *Reprod Toxicol.* 2010;30:566–572.

9. Utsunomiya T, Tanaka T, Utsunomiya H, Umesaki N. A novel molecular mechanism for anticancer drug-induced ovarian failure: Irinotecan HCl, an anticancer topoisomerase I inhibitor, induces specific FasL expression in granulosa cells of large ovarian follicles to enhance follicular apoptosis. *Int J Oncol.* 2008;32:991–1000.

10. Depalo R, Nappi L, Loverro G, et al. Evidence of apoptosis in human primordial and primary follicles. *Hum Reprod.* 2003;18:2678–2682.

11. Morita Y, Perez GI, Paris F, et al. Oocyte apoptosis is suppressed by disruption of the acid sphingomyelinase gene or by sphingosine-1-phosphate therapy. *Nat Med.* 2000;6:1109–1114.

12. Morita Y, Perez GI, Maravei DV, Tilly KI, Tilly JL. Targeted expression of Bcl-2 in mouse oocytes inhibits ovarian follicle atresia and prevents spontaneous and chemotherapy-induced oocyte apoptosis *in vitro. Mol Endocrinol.* 1999;13:841–850.

13. Bergeron L, Perez GI, Macdonald G, et al. Defects in regulation of apoptosis in caspase-2-deficient mice. *Genes Dev.* 1998;12:1304–1314.

14. Tilly JL. Molecular and genetic basis of normal and toxicant-induced apoptosis in female germ cells. *Toxicol Lett.* 1998;102-103:497–501.

15. Takai Y, Matikainen T, Jurisicova A, et al. Caspase-12 compensates for lack of caspase-2 and caspase-3 in female germ cells. *Apoptosis.* 2007;12:791–800.

16. Kurita T, Cunha GR, Robboy SJ, Mills AA, Medina RT. Differential expression of p63 isoforms in female reproductive organs. *Mech Dev.* 2005;122:1043–1055.

17. Suh EK, Yang A, Kettenbach A, et al. p63 protects the female germ line during meiotic arrest. *Nature.* 2006;444:624–628.

18. Livera G, Petre-Lazar B, Guerquin MJ, Trautmann E, Coffigny H, Habert R. p63 null mutation protects mouse oocytes from radio-induced apoptosis. *Reproduction.* 2008;135:3–12.

19. Gonfloni S, Di Tella L, Caldarola S, et al. Inhibition of the c-Abl-TAp63 pathway protects mouse oocytes from chemotherapy-induced death. *Nat Med.* 2009;15:1179–1185.

20. Kim SY, Cordeiro MH, Serna VA, et al. Rescue of platinum-damaged oocytes from programmed cell death through inactivation of the p53 family signaling network. *Cell Death Differ.* 2013;20:987–997.

21. Kerr JB, Hutt KJ, Michalak EM, et al. DNA damage-induced primordial follicle oocyte apoptosis and loss of fertility require TAp63-mediated induction of Puma and Noxa. *Mol Cell.* 2012;48(3):343–352.

22. Albanese R. Induction and transmission of chemically induced chromosome aberrations in female germ cells. *Environ Mol Mutagen.* 1987;10:231–243.

23. Yoshida K, Miki Y. Enabling death by the Abl tyrosine kinase: mechanisms for nuclear shuttling of c-Abl in response to DNA damage. *Cell Cycle.* 2005;4:777–779.

24. Kharbanda S, Pandey P, Morris PL, et al. Functional role for the c-Abl tyrosine kinase in meiosis I. *Oncogene.* 1998;16:1773–1777.

25. Kharbanda S, Yuan ZM, Weichselbaum R, Kufe D. Determination of cell fate by c-Abl activation in the response to DNA damage. *Oncogene.* 1998;17:3309–3318.

26. Kerr JB, Hutt KJ, Cook M, et al. Cisplatin-induced primordial follicle oocyte killing and loss of fertility are not prevented by imatinib. *Nat Med.* 2012;18:1170–1172: author reply 1172–1174.

27. Maiani E, Di Bartolomeo C, Klinger FG, et al. Reply to: Cisplatin-induced primordial follicle oocyte killing and loss of fertility are not prevented by imatinib. *Nat Med.* 2012;18:1172–1174.

28. Morgan S, Lopes F, Gourley C, Anderson RA, Spears N. Cisplatin and doxorubicin induce distinct mechanisms of ovarian follicle loss; imatinib provides selective protection only against cisplatin. *PLoS One.* 2013;8:e70117.

29. Sedletska Y, Giraud-Panis MJ, Malinge JM. Cisplatin is a DNA-damaging antitumour compound triggering multifactorial biochemical responses in cancer cells: importance of apoptotic pathways. *Curr Med Chem Anticancer Agents.* 2005;5:251–265.

30. Leong CO, Vidnovic N, DeYoung MP, Sgroi D, Ellisen LW. The p63/p73 network mediates chemosensitivity to cisplatin in a biologically defined subset of primary breast cancers. *J Clin Invest.* 2007;117:1370–1380.

31. Soleimani R, Heytens E, Oktay K. Enhancement of neoangiogenesis and follicle survival by sphingosine-1-phosphate in human ovarian tissue xenotransplants. *PLoS One.* 2011;6:e19475.

32. Kaya H, Desdicioglu R, Sezik M, et al. Does sphingosine-1-phosphate have a protective effect on cyclophosphamide- and irradiation-induced ovarian damage in the rat model? *Fertil Steril.* 2008;89:732–735.

33. Zelinski MB, Murphy MK, Lawson MS, et al. *In vivo* delivery of FTY720 prevents radiation-induced ovarian failure and infertility in adult female nonhuman primates. *Fertil Steril.* 2011;95:1440–5, e1–7.

34. Hancke K, Strauch O, Kissel C, Gobel H, Schafer W, Denschlag D. Sphingosine 1-phosphate protects ovaries from chemotherapy-induced damage *in vivo. Fertil Steril.* 2007;87:172–177.

35. Li F, Turan V, Lierman S, Cuvelier C, De Sutter P, Oktay K. Sphingosine-1-phosphate prevents chemotherapy-induced human primordial follicle death. *Hum Reprod.* 2014;29:107–113.

36. Paris F, Perez GI, Fuks Z, et al. Sphingosine 1-phosphate preserves fertility in irradiated female mice without propagating genomic damage in offspring. *Nat Med.* 2002;8:901–902.

37. Skaznik-Wikiel ME, McGuire MM, Sukhwani M, et al. Granulocyte colony-stimulating factor with or without stem cell factor extends time to premature ovarian insufficiency in female mice treated with alkylating chemotherapy. *Fertil Steril.* 2013;99:2045–54, e3.

38. Meirow D, Dor J, Kaufman B, et al. Cortical fibrosis and blood-vessels damage in human ovaries exposed to chemotherapy. Potential mechanisms of ovarian injury. *Hum Reprod.* 2007;22:1626–1633.

39. Solaroglu I, Cahill J, Jadhav V, Zhang JH. A novel neuroprotectant granulocyte-colony stimulating factor. *Stroke.* 2006;37:1123–1128.

40. Pastuszko P, Schears GJ, Pirzadeh A, et al. Effect of granulocyte-colony stimulating factor on expression of selected proteins involved in regulation of apoptosis in the brain of newborn piglets after cardiopulmonary bypass and deep hypothermic circulatory arrest. *J Thorac Cardiovasc Surg.* 2012;143:1436–1442.

41. Laoag-Fernandez JB, Matsuo H, Murakoshi H, Hamada AL, Tsang BK, Maruo T. 3,5,3′-Triiodothyronine down-regulates Fas and Fas ligand expression and suppresses caspase-3 and poly (adenosine 5′-diphosphate-ribose) polymerase cleavage and apoptosis in early placental extravillous trophoblasts *in vitro. J Clin Endocrinol Metab.* 2004;89:4069–4077.

42. Sukocheva OA, Carpenter DO. Anti-apoptotic effects of 3,5,3′-tri-iodothyronine in mouse hepatocytes. *J Endocrinol.* 2006;191:447–458.

43. Verga Falzacappa C, Panacchia L, Bucci B, et al. 3,5,3′-triiodothyronine (T3) is a survival factor for pancreatic beta-cells undergoing apoptosis. *J Cell Physiol.* 2006;206:309–321.

44. Verga Falzacappa C, Mangialardo C, Patriarca V, et al. Thyroid hormones induce cell proliferation and survival in ovarian granulosa cells COV434. *J Cell Physiol.* 2009;221:242–253.

45. Verga Falzacappa C, Timperi E, Bucci B, et al. T(3) preserves ovarian granulosa cells from chemotherapy-induced apoptosis. *J Endocrinol*. 2012;215:281–289.

46. Zhang C, Xia G, Tsang BK. Interactions of thyroid hormone and FSH in the regulation of rat granulosa cell apoptosis. In: Elite, ed. *Front Biosci*. 3. 2011:1401–1413.

47. Ting AY, Petroff BK. Tamoxifen decreases ovarian follicular loss from experimental toxicant DMBA and chemotherapy agents cyclophosphamide and doxorubicin in the rat. *J Assist Reprod Genet*. 2010;27:591–597.

48. Fornier MN, Modi S, Panageas KS, Norton L, Hudis C. Incidence of chemotherapy-induced, long-term amenorrhea in patients with breast carcinoma age 40 years and younger after adjuvant anthracycline and taxane. *Cancer*. 2005;104:1575–1579.

49. Yoo C, Yun MR, Ahn JH, et al. Chemotherapy-induced amenorrhea, menopause-specific quality of life, and endocrine profiles in premenopausal women with breast cancer who received adjuvant anthracycline-based chemotherapy: a prospective cohort study. *Cancer Chemother Pharmacol*. 2013;72:565–575.

50. Swain SM, Land SR, Ritter MW, et al. Amenorrhea in premenopausal women on the doxorubicin-and-cyclophosphamide-followed-by-docetaxel arm of NSABP B-30 trial. *Breast Cancer Res Treat*. 2009;113:315–320.

51. Partridge A, Gelber S, Gelber RD, Castiglione-Gertsch M, Goldhirsch A, Winer E. Age of menopause among women who remain premenopausal following treatment for early breast cancer: long-term results from International Breast Cancer Study Group Trials V and VI. *Eur J Cancer*. 2007;43:1646–1653.

52. Perez-Fidalgo JA, Rosello S, Garcia-Garre E, et al. Incidence of chemotherapy-induced amenorrhea in hormone-sensitive breast cancer patients: the impact of addition of taxanes to anthracycline-based regimens. *Breast Cancer Res Treat*. 2010;120:245–251.

53. Valentini A, Finch A, Lubinski J, et al. Chemotherapy-induced amenorrhea in patients with breast cancer with a *BRCA1* or *BRCA2* mutation. *J Clin Oncol*. 2013;31:3914–3919.

54. Jung M, Shin HJ, Rha SY, et al. The clinical outcome of chemotherapy-induced amenorrhea in premenopausal young patients with breast cancer with long-term follow-up. *Ann Surg Oncol*. 2010;17:3259–3268.

55. Berliere M, Duhoux FP, Dalenc F, et al. Tamoxifen and ovarian function. *PLoS One*. 2013;8:e66616.

56. Dubey RK, Tyurina YY, Tyurin VA, et al. Estrogen and tamoxifen metabolites protect smooth muscle cell membrane phospholipids against peroxidation and inhibit cell growth. *Circ Res*. 1999;84:229–239.

57. Nathan L, Chaudhuri G. Antioxidant and prooxidant actions of estrogens: potential physiological and clinical implications. *Semin Reprod Endocrinol*. 1998;16:309–314.

58. Mahran YF, El-Demerdash E, Nada AS, Ali AA, Abdel-Naim AB. Insights into the protective mechanisms of tamoxifen in radiotherapy-induced ovarian follicular loss: impact on insulin-like growth factor 1. *Endocrinology*. 2013;154:3888–3899.

59. Zhou J, Kumar TR, Matzuk MM, Bondy C. Insulin-like growth factor I regulates gonadotropin responsiveness in the murine ovary. *Mol Endocrinol*. 1997;11:1924–1933.

60. Baeza I, Fdez-Tresguerres J, Ariznavarreta C, De la Fuente M. Effects of growth hormone, melatonin, oestrogens and phytoestrogens on the oxidized glutathione (GSSG)/reduced glutathione (GSH) ratio and lipid peroxidation in aged ovariectomized rats. *Biogerontology*. 2010;11:687–701.

61. Armstrong DG, Webb R. Ovarian follicular dominance: the role of intraovarian growth factors and novel proteins. *Rev Reprod*. 1997;2:139–146.

62. Rabinovici J, Jaffe RB. Development and regulation of growth and differentiated function in human and subhuman primate fetal gonads. *Endocr Rev.* 1990;11:532–557.

63. Hsueh AJ, McGee EA, Hayashi M, Hsu SY. Hormonal regulation of early follicle development in the rat ovary. *Mol Cell Endocrinol.* 2000;163:95–100.

64. Kalich-Philosoph L, Roness H, Carmely A, et al. Cyclophosphamide triggers follicle activation causing ovarian reserve 'burn out'; AS101 prevents follicle loss and preserves fertility. *Sci Transl Med.* 2013;5(185):185ra62.

65. Adhikari D, Liu K. Molecular mechanisms underlying the activation of mammalian primordial follicles. *Endocr Rev.* 2009;30:438–464.

66. Kawamura K, Cheng Y, Suzuki N, et al. Hippo signaling disruption and Akt stimulation of ovarian follicles for infertility treatment. *Proc Natl Acad Sci USA.* 2013;110:17474–17479.

67. Li J, Kawamura K, Cheng Y, et al. Activation of dormant ovarian follicles to generate mature eggs. *Proc Natl Acad Sci USA.* 2010;107:10280–10284.

68. McLaughlin M, Kinnell HL, Anderson RA, Telfer EE. Inhibition of phosphatase and tensin homologue (PTEN) in human ovary *in vitro* results in increased activation of primordial follicles but compromises development of growing follicles. *Mol Hum Reprod.* 2014;20:736–744.

69. Roness, H. ESHRE, London; 2013.

70. Sobinoff AP, Nixon B, Roman SD, McLaughlin EA. Staying alive: PI3K pathway promotes primordial follicle activation and survival in response to 3-MC induced ovotoxicity. *Toxicol Sci.* 2012;128(1):258–271.

71. Sredni B. Immunomodulating tellurium compounds as anti-cancer agents. *Semin Cancer Biol.* 2012;22:60–69.

72. Makarovsky D, Kalechman Y, Sonino T, et al. Tellurium compound AS101 induces PC12 differentiation and rescues the neurons from apoptotic death. *Ann NY Acad Sci.* 2003;1010:659–666.

73. Kalechman Y, Albeck M, Oron M, et al. Protective and restorative role of AS101 in combination with chemotherapy. *Cancer Res.* 1991;51:1499–1503.

74. Kalechman Y, Shani A, Sotnik-Barkai I, Albeck M, Sredni B. The protective role of ammonium trichloro(dioxoethylene-O,O′)tellurate in combination with several cytotoxic drugs acting by different mechanisms of action. *Cancer Res.* 1993;53:5962–5969.

75. Kalechman Y, Sotnik-Barkai I, Albeck M, Sredni B. Protection of bone marrow stromal cells from the toxic effects of cyclophosphamide *in vivo* and of ASTA-Z 7557 and etoposide *in vitro* by ammonium trichloro(dioxyethylene-O-O′)tellurate (AS101). *Cancer Res.* 1993;53:1838–1844.

76. Sredni B, Albeck M, Kazimirsky G, Shalit F. The immunomodulator AS101 administered orally as a chemoprotective and radioprotective agent. *Int J Immunopharmacol.* 1992;14:613–619.

77. Kiyomiya K, Matsuo S, Kurebe M. Mechanism of specific nuclear transport of adriamycin: the mode of nuclear translocation of adriamycin-proteasome complex. *Cancer Res.* 2001;61:2467–2471.

78. Roti Roti EC, Ringelstetter AK, Kropp J, Abbott DH, Salih SM. Bortezomib prevents acute Doxorubicin ovarian insult and follicle demise, improving the fertility window and pup birth weight in mice. *PLoS One.* 2014;9:e108174.

79. Fojo AT, Ueda K, Slamon DJ, Poplack DG, Gottesman MM, Pastan I. Expression of a multidrug-resistance gene in human tumors and tissues. *Proc Natl Acad Sci USA.* 1987;84:265–269.

80. Lepper ER, Nooter K, Verweij J, Acharya MR, Figg WD, Sparreboom A. Mechanisms of resistance to anticancer drugs: the role of the polymorphic ABC transporters ABCB1 and ABCG2. *Pharmacogenomics.* 2005;6:115−138.

81. Salih SM. Retrovirus-mediated multidrug resistance gene (MDR1) overexpression inhibits chemotherapy-induced toxicity of granulosa cells. *Fertil Steril.* 2011;95:1390−1396, e1−6.

82. Brayboy LM, Oulhen N, Witmyer J, Robins J, Carson S, Wessel GM. Multidrug-resistant transport activity protects oocytes from chemotherapeutic agents and changes during oocyte maturation. *Fertil Steril.* 2013;100:1428−1435.

83. Tahover E, Patil YP, Gabizon AA. Emerging delivery systems to reduce doxorubicin cardiotoxicity and improve therapeutic index: focus on liposomes. *Anticancer Drugs.* 2014.

84. Ahn RW, Barrett SL, Raja MR, et al. Nano-encapsulation of arsenic trioxide enhances efficacy against murine lymphoma model while minimizing its impact on ovarian reserve *in vitro* and *in vivo*. *PLoS One.* 2013;8:e58491.

85. Jurisicova A, Lee HJ, D'Estaing SG, Tilly J, Perez GI. Molecular requirements for doxorubicin-mediated death in murine oocytes. *Cell Death Differ.* 2006;13:1466−1474.

86. Akdemir A, Zeybek B, Akman L, et al. Granulocyte-colony stimulating factor decreases the extent of ovarian damage caused by cisplatin in an experimental rat model. *J Gynecol Oncol.* 2014;25(4):328−333.

87. Sverrisdottir A, Nystedt M, Johansson H, Fornander T. Adjuvant goserelin and ovarian preservation in chemotherapy treated patients with early breast cancer: results from a randomized trial. *Breast Cancer Res Treat.* 2009;117:561−567.